A colour atlas of
Histological
Staining Techniques

A colour atlas of
Histological Staining Techniques

ARTHUR SMITH

FIMLS, LI.Biol
Senior Chief Technologist
Department of Oral Pathology
Institute of Dental Surgery
Eastman Dental Hospital, London

JOHN BRUTON

AIST
Senior Technologist
Department of Oral Pathology
Institute of Dental Surgery
Eastman Dental Hospital, London

WOLFE MEDICAL PUBLICATIONS LTD
10 Earlham Street   London WC2

Copyright © Arthur Smith and John Bruton 1977
Published by Wolfe Medical Publications Ltd 1977
Printed by Sackville Press Billericay Ltd
ISBN 07234 0721 5

General Editor, Wolfe Medical Books
G Barry Carruthers MD (Lond)

Other books in this series already published
A colour atlas of Haematological Cytology
A colour atlas of General Pathology
A colour atlas of Oro-Facial Diseases
A colour atlas of Ophthalmological Diagnosis
A colour atlas of Renal Diseases
A colour atlas of Venereology
A colour atlas of Dermatology
A colour atlas of Infectious Diseases
A colour atlas of Ear, Nose & Throat Diagnosis
A colour atlas of Rheumatology
A colour atlas of Microbiology
A colour atlas of Forensic Pathology
A colour atlas of Paediatrics
A colour atlas of Histology
A colour atlas of General Surgical Diagnosis
A colour atlas of Physical Signs in General Medicine
A colour atlas of Tropical Medicine and Parasitology
A colour atlas of Cardiac Pathology

A colour atlas of Human Anatomy

Further titles now in preparation
A colour atlas of Oral Anatomy
A colour atlas of Neuropathology
A colour atlas of Diabetes Mellitus
A colour atlas of the Pathology of Lymph Nodes
A colour atlas of Standard Gynaecological Operations
A colour atlas of Oral Medicine
An atlas of Cardiology: ECGs and Chest X-Rays

CONTENTS

To Ruth Lynne,
Andrew John,
Kevin and Daniel

PREFACE

Histological 'staining' technique is the art, craft or science of demonstrating the various tissue components by differential dyeing or metallic coating. The majority of 'staining' techniques used today were conceived by accident: they were observed by the perceptive eye or resulted from inspired guesswork. Frequently the precise chemical constituents of the tissue were, and are, unknown and the 'theory' of the staining mechanism empirical. Techniques based on the known chemistry of the tissue reacting to a known and understood chemical reaction are few in number. In the not too distant future this ratio may change for the specialist. But for the student and less specialised histologist, techniques which are easy to use under non-research conditions will be required for the foreseeable future.

This atlas provides a representative collection of working instructions for histological 'staining' techniques combined with the illustrations of the end staining results which the student, postgraduate student and 'worker' in a non-specialised histology laboratory are likely to need.

It is not our intention to instruct the experienced histologist, but we hope that he will find the book useful in a specialist laboratory when he needs to show students and research workers how a particular 'common' tissue element, normally outside that laboratory's orbit, can be demonstrated. It will also serve as a memory aid to those who are interested in histology.

This atlas is *not* a substitute for a textbook, and the student seeking knowledge in depth must refer to one of these. It is a visual introduction for the beginner and the inexperienced to an aesthetically pleasing subject, an easy reference manual for the workbench.

ACKNOWLEDGEMENTS

In all aspects of life the individual learns something great or small from every contact with others, and this is true of contact with professional colleagues.

While it is impracticable to record the names of the many who have helped us to learn the 'craft' we ask them when reading this acknowledgement to accept our individual appreciation. We record our indebtedness to Professor I. R. H. Kramer for encouraging us to experiment with techniques and then to communicate our results. To Miss Sharon Wilkinson must go the credit for translating the awful ink scribbles of one of us into clear type. It is doubtful whether anybody else could have done so. The publishers have remained cool, calm, confident and always constructive, never obstructive. Finally we offer our apologies and thanks to our families, who must have suffered when the going was tougher than usual.

INTRODUCTION

In 1902 Gustav Mann wrote a book called *Physiological Histology* which dealt with a subject then still in its infancy. The following quotation from this book sums up the position as he saw it then, and how he might well see it to a similar extent today.

'The method of staining, once having taken root in the animal histologist, grew and grew, till to be an histologist became practically synonymous with being a dyer with this difference, that the professional dyer knew what he was about, while the histologist with few exceptions did not know, nor does he to the present day.'

The perfect histological technique would produce results which were without exception specific and in which the chance of human error was absent. In practice it must be accepted that few techniques are specific and the chance of human error is ever present. We must therefore accept a degree of compromise between the ideal and the practical. However, while this compromise is necessary, we must never accept a compromise in the interpretation of the staining results or by ignoring possible human error when applying the technique, particularly where differentiation is concerned.

To confirm the staining results of a particular technique a known control section should be treated simultaneously, and if possible a second technique should be applied to a companion section as a precaution until experience makes this second precaution less necessary. Nevertheless, however experienced one is with a 'special' staining technique a control section should without exception be treated simultaneously to check the solutions and the absence of human error. This procedure is of prime importance in clinical diagnoses or research procedures.

With the application of histological technique and interpretation of the staining results there is no substitute for knowledge of normal histology, experience of 'doing' the technique and 'reading' the finished preparation, all of which must be carried out with integrity.

The principal difficulty is the presence in the same section of many different tissue structures that have some degree of affinity for any one dye or metallic impregnation, thus causing difficulties in demonstrating a single substance or structure. Frequently the only way to achieve this is by the most carefully controlled differentiation, with the risk of producing a false negative or positive result.

Why does this difficulty exist? The chemistry of the fixed tissue, or even perhaps the fresh tissue, is poorly understood and although the majority of techniques work, they are empirical. They have been used for a long time without too much questioning or with no attempt to replace them with methods based on scientific formulae. So the techniques in use today range from a few histochemical techniques where the reaction equations are known, through others where inspired calculated guesses are advanced as to the mechanics of their working, to others where the rationale is completely unknown. Since the majority of techniques are empirical, staining results should always be viewed with reasonable suspicion and if the slightest doubt exists, checked and cross-checked.

Provided that the dye sample has been checked for suitability the most common cause of failure is the human element. Frequently, the error is due to lack of experience or a mental aberration due to absence of concentration.

To correct the first requires patience and practice with close attention to detail. Correction of the second is more difficult: familiarity and pressure of work must not be allowed to obscure the need for revision and the adequate checking of procedures. For example, most techniques require that the sections at some stage after staining shall come into contact with water, and it is surprisingly easy to overlook the importance of avoiding all contact with water after staining in, for example, Best's Carmine, unless revision with comprehension of the whole technique has taken place and there has not been simply a superficial glance at the technique to check staining times.

The composite question most frequently asked is 'How long should the section be in the staining solution; what length of time is required in the differentiator; and is this section differentiated correctly and are these colours right?' There is no single easy answer. The answer to each part depends a great deal on the individual's training, acuteness of colour perception and the colour temperature of the light source associated with the microscope.

With most techniques the staining times given have been found to give the desired results, but they can be, and are, varied without detriment to the end staining results. Experience and familiarity with the technique will teach the permissible variations.

Differentiation of sections almost invariably requires frequent examination of the section under the microscope, followed by thorough removal of the differentiator. The degree of differentiation presents more of a problem. With a number of techniques the finished results can vary from 'heavy' (under-differentiated to some) or 'light' (over-differentiated to others), with varying densities in between, personal preference or a long-accepted local standard indicating a 'correct' result. Advice on this issue can only be that the tissue elements required to be demonstrated must be clearly and unmistakably observable; they must not be obscured by being overstained or lost through over-differentiation. The use of a simultaneously stained control section will considerably assist in eliminating this problem.

Individuality is most marked in the acceptance of variations in colour density and the shade that is accepted as 'correct'.

Haematoxylin and eosin sequences provide a good example of the difficulties in stating 'correct' staining times and the correctly stained preparation. In general the criteria for a good H & E stained preparation are that the haematoxylin should be eliminated from tissues where it is not required, but present clearly where it is required, and that the eosin should be clear pink-red, not purplish. Reference

11

to Figure 2 illustrates the problem of stating dogmatically that a precise timing is correct and that this is the correct picture.

Defining correct colour is also a problem. If you look at a colour chart (Figure 1) you may find it hard to decide when 'red' is red and not orange or purplish, or for that matter when 'green' is green and not blue or yellow. Our colour perception is influenced by our degree of artistic training and the labels we have been taught to apply to our interpretations of a colour or shade. A further complication is the great variability in the light sources used for microscope illumination. A source of light of an incorrect colour temperature alters the shade of colour. Only familiarity and experience with a wide variety of techniques will teach what colour or shade is acceptable.

To attempt histological technique without at least a working knowledge of normal histology is to risk, at the best, unreliable results and at the worst complete failure. If you do not know the principal tissues that should be present in the section it is not possible to demonstrate with any degree of certainty unless the technique is specific; and it must be conceded that few techniques are specific. For example, it is not at all difficult to reverse the colours in Masson's trichome so that the muscle in a section is stained blue. The elementary knowledge that heart wall is predominantly muscle will avoid the error of trying to produce a mainly blue section. By the same token a section composed mainly of collagen stained by the same technique should appear blue and not red.

A knowledge of the micro-anatomical relationship of tissues to one another and an understanding of cell appearances will greatly assist in the carrying out of a staining technique and the 'reading' of the resulting stained preparation will be more rewarding and accurate.

The thickness at which sections are cut influences the usefulness of the sections. As a generalisation, thin sections are required for observing cellular detail and thick sections for tracing the ramifications of fibres. For general usefulness, sections 4 to 5 micrometers are almost invariably advocated. Unless specifically stated in the photomicrograph legend or detailed technique, all sections photomicrographed in this atlas were cut at a setting of 5 micrometers.

The aim of fixation is to preserve the cells and tissues in as lifelike a condition as possible. The effect of fixation is to prevent autolysis and putrefaction of the tissues and to render them resistant to damage by osmotic pressure. In some cases a fixative also acts as a mordant for a particular staining reaction. Ideally, the choice of a fixative should be governed by the tissue element it is desired to demonstrate, but in practice this ideal is seldom obtainable, because factors other than fixation must be considered.

First, when dealing with human material, it is rarely possible to arrange for the fresh specimen to be sent immediately to the laboratory. For this reason the specimen is placed in a container of fixative within the environs of the operating theatre or surgery.

As this procedure is performed by non-technical staff, the technique should consist of nothing more than the immersion of the specimen in the fixative and its despatch to the laboratory as soon as possible. Also, it is preferable that the fixative should be non-corrosive, for a corrosive fluid might prove dangerous when used under these conditions.

Second, the time that tissue may remain in many fixatives is critical and this restriction may make these fixatives unsuitable for routine use as they do not allow for unavoidable delays which the specimen may undergo before reaching the laboratory.

Third, some fixatives penetrate tissue poorly; therefore, only thin slices of tissue may be fixed in them. This group of fixatives is more applicable to research procedures than to routine use.

Fourth, it may well be that part of the specimen may be required for museum mounting or storage as a wet teaching specimen. Therefore, it is essential that the fixative must be compatible with these eventualities. Furthermore, compared to soft tissue, calcified tissues are relatively impermeable to the fixatives. To allow for this the fixative must be continuous in action over a prolonged period.

The specimen must invariably be fixed *in toto* to avoid damaging the tissues. It follows that the complete specimen can be fixed in only one fixative, the alternative being to cut the specimen into a number of blocks so that a variety of fixatives may be used. To use this alternative is to run the risk of destroying the tissue relationship and causing other artefacts.

Bearing in mind these requirements, it is easy to appreciate why formol saline has become the general-purpose fixative. Walker Hall and Herxheimer state that formalin was first introduced into histological technique by F. Blum in 1893. Since then numerous authors have recommended, with varying degrees of enthusiasm, its use for general-purpose fixation. However, in few instances have their recommendations been unqualified, for the use of formalin has certain disadvantages. Perhaps the situation has best been summed up by Everson Pearse (1953): 'It is better to become reconciled to the evils of formalin.' He also writes that 'from the histochemical, as well as the histological point of view, the objection to soft fixation can be overcome by the freezing method.'

Formol saline has the following advantages and disadvantages:

Advantages
1 Rapid (or slow) and even penetration of the tissues.
2 Specimens may be fixed *in toto,* even large ones, although with these it would be better to perfuse or slice them. The specimen may also be stored in this fixative.
3 This fluid is suitable for use in the environs of the operating theatre or surgery.
4 Penetration is continuous (Underhill 1932) but there is little danger of overfixation.
5 Fat is preserved.
6 Permits the use of most staining and impregnation techniques, including special methods for nerves.
7 Specimens may be processed for museum mounting after fixation in this fluid (Smith 1959).

Disadvantages
1 Owing to 'soft fixation' there may be considerable shrinkage of the tissue during subsequent embedding by the paraffin method, and to a lesser extent by celloidin embedding.
2 Preservation of the nuclear detail is inferior to some nuclear fixatives.
3 Formalin pigment may form in the tissue (easily recognised and removed).
4 Formol saline is liable to cause dermatitis and is also irritating to

14

the nasal mucosa (rubber gloves should be worn and the room well ventilated).

Experience has shown that it is frequently difficult to use the 'correct' fixative as recommended by the technique's author, but fortunately it has been shown that seldom is this recommendation essential for good staining results, with the exception of the tissue elements that require specialised fixation for their preservation.

Bearing in mind the foregoing, *almost all the illustrations in this atlas are of tissue fixed in 5 per cent formol saline and processed by the technique detailed,* unless otherwise stated.

A sobering thought for histologists should be that their finished histological preparations are a record of their skill and expertise for as long as their preparations exist. For the section to be of maximum use the cutting should be of a high standard at the most suitable thickness and the staining must clearly show what is intended to be seen. Where possible the mounted preparation should be dried and the surplus mountant cleaned off. Finally the section must be clearly labelled.

If this standard is aimed for, the preparation will be of maximum use and a pleasure to exhibit; otherwise it will be an object of limited use and repeated apology.

The illustrations in this atlas are photomicrographs of tissues usually available. No attempt has been made to find material which is rarely available to the general worker. The criterion used in selecting the illustrations was the clarity with which a particular tissue structure could be demonstrated and compared with other structures in the selected technique.

In the legends to the photomicrograph, the naming of the tissue has been deliberately omitted; only the tissue elements demonstrated by the technique are listed. It is important that the student learns to find the tissue elements where they exist and to avoid the erroneous idea that the tissue quoted by the authors is the best or only site for the tissue element(s) demonstrated. For example, collagen fibres are demonstrated as such by the appropriate technique, irrespective of which part of the animal is selected for sectioning. Obviously, some specialised tissue elements are to be found in a particular site, hence

15

the authors' previous injunction that the student has at least a working knowledge of normal histology.

The selection of techniques for inclusion in this atlas has been decided by experience and the knowledge that they 'work'. To illustrate all the available histological techniques together with their minor modifications would be the work of many expensive volumes. The authors' objective is to stimulate interest, to encourage the reader to practise, to seek knowledge in greater depth from original papers, early textbooks and current publications and, equally important, to enjoy attempting to seek the answers to the many questions associated with histological 'staining' techniques.

COLOUR CHART (Figure 1)

PART 1

**Photomicrographs illustrating
end staining results of detailed techniques**

MAYER'S HAEMALUM

2. This haematoxylin is used progressively; differentiation not normally required. Which staining time is correct? Personal preference for 'heavy or light' staining, colour perception, quality of microscope, colour temperature of light source and personal experience are all factors influencing choice. However, the tissue element to be examined must be clearly demonstrated and the staining 'artefact' standardised if the technique is to have any value.

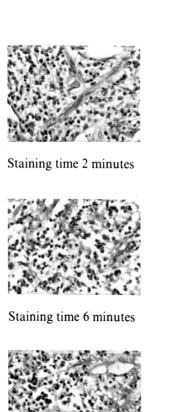

Staining time 2 minutes

Staining time 4 minutes

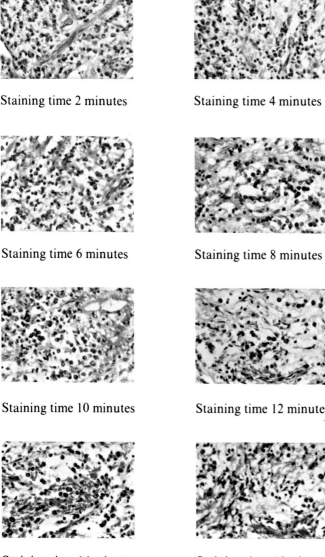

Staining time 6 minutes

Staining time 8 minutes

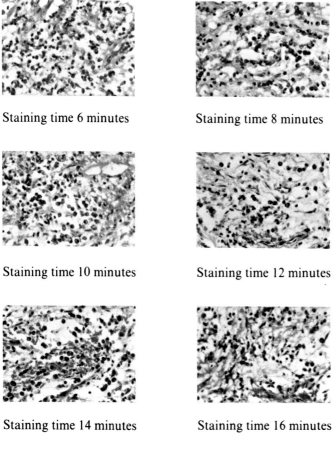

Staining time 10 minutes

Staining time 12 minutes

Staining time 14 minutes

Staining time 16 minutes

CALCIFIED TISSUE

Which haematoxylin to use for staining decalcified tissue? For demonstrating the structure of this tissue Ehrlich's or Cole's haematoxylin are preferred, remembering that acid decalcification inhibits haematoxylin staining but enhances eosin staining.

3. Ehrlich's acid haematoxylin and eosin (page 122)
× 100

Nuclei: blue-black
Cartilage: bluish
Incremental lines: blue
Cytoplasm ⎫ pink
Connective ⎬ to
tissue fibres ⎭ red

4. Cole's haematoxylin and eosin (page 121)
× 100

Nuclei: blue-black
Cartilage: bluish
Incremental lines: blue
Cytoplasm ⎫ pink
Connective ⎬ to
tissue fibres ⎭ red

5. Harris's haematoxylin and eosin (page 133)
× 100

Nuclei: blue-black
Cartilage: bluish
Incremental lines: blue
Cytoplasm ⎫ pink
Connective ⎬ to
tissue fibres ⎭ red

6. Mayer's haematoxylin and eosin (page 145)
× 100

Nuclei: blue-black
Cartilage: bluish
Incremental lines: blue
Cytoplasm ⎫ pink
Connective ⎬ to
tissue fibres ⎭ red

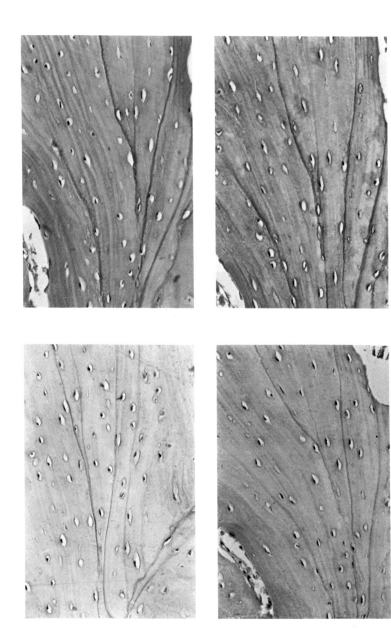

23

CALCIFIED TISSUE

7. van Gieson's picro fuchsin (page 167)
× 200

Collagen: ranging from pink to red
Muscle and erythrocytes: yellow

It is an advantage not to counterstain decalcified tissue
with haematoxylin when using van Gieson. The collagen
fibres appear to stain with differing intensities of red, single
fibres appearing pink and dense bundles deep red. This is
the easiest technique for collagen and is adequate for most
purposes.

8. Masson's trichrome (blue or green) (page 143)
× 25

Collagen: blue or green
Denatured collagen: orange-yellow
Cytoplasm: red
Mucin: blue or green
Muscle: red

This section illustrates that normal collagen stains
selectively from that denatured by disease, as is the case of
caries in this tooth section.

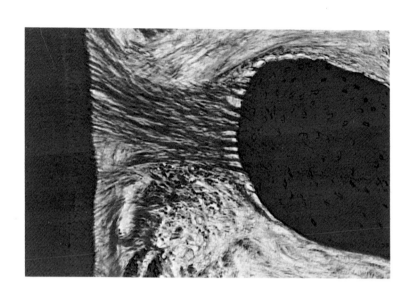

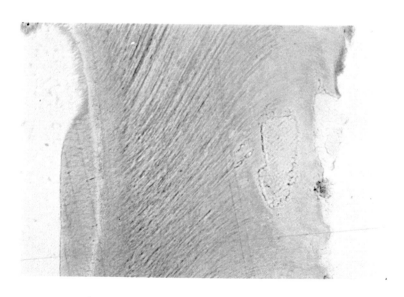

CALCIFIED TISSUE

9. Schmorl's picro thionin (page 157)
× 100

Lacunae
Canaliculi
Dentinal tubules } dark brown
Lateral branches
Background: yellow to yellow-brown

This technique is not a tissue-staining method but a method of precipitating picrates into spaces within calcified tissue. It may be applied to ground sections of un-decalcified tissue or to sections of decalcified tissue.

10. Schmorl's picro thionin (page 157)
× 100

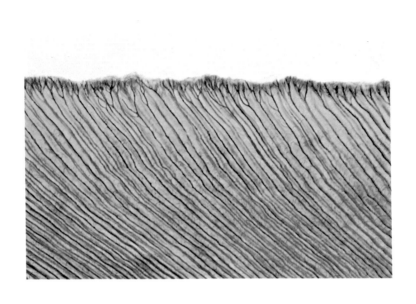

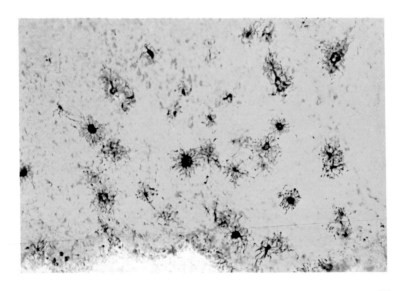

CALCIUM

11. Ehrlich's haematoxylin and eosin (page 122)
× 25

Calcium deposits: purplish

12. Von Kossa's silver method (page 170)
× 25

Calcium deposits: dark brown black
Nuclei: red

13. Alizarin red (page 113)
× 25

Calcium deposits: orange-red

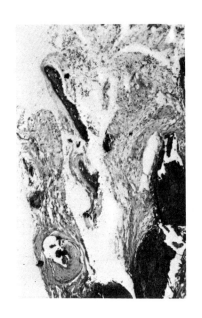

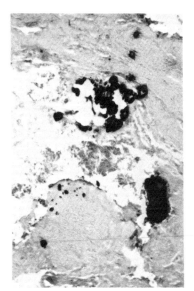

29

CARTILAGE

**14. Ehrlich's
haematoxylin and eosin**
(page 122)
$\times$ 100

15. Metachromasia
(page 146)
$\times$ 100

Cartilage: purple

16. Periodic acid Schiff
(page 152)
$\times$ 100

Cartilage: reddish

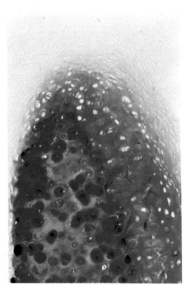

31

CONNECTIVE TISSUE

Basement membrane

**17. Ehrlich's
haematoxylin and eosin**
(page 122)
× 100

**18. Silver impregnation.
Jones' (counterstain
with eosin)** (page 159)
× 100

Basement membrane:
black
Other tissue: pink to
red

19. Periodic acid Schiff
(page 152)
× 100

Positive material (base-
ment membrane):
reddish-purple
Nuclei: faint grey

This technique demon-
strates many tissue sub-
stances and structures
in the same tintorial
tint; interpretation
therefore frequently
depends on
morphology.

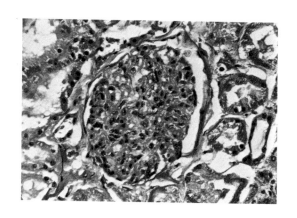

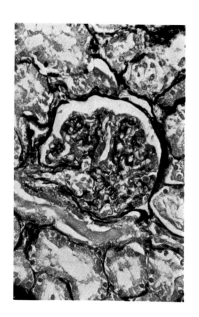

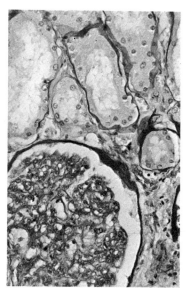

Collagen

20. Ehrlich's haematoxylin and eosin (page 122)
× 100

Collagen and other connective tissues: pink to red
Nuclei: blue

21. van Gieson's picro fuchsin and iron haematoxylin
(page 167)
× 100

Collagen: ranging from pink to red
Muscle and erythrocytes: yellow
Nuclei: blackish

This is the easiest collagen technique and is adequate for
most purposes although personal preference may dictate
the use of a different colour.

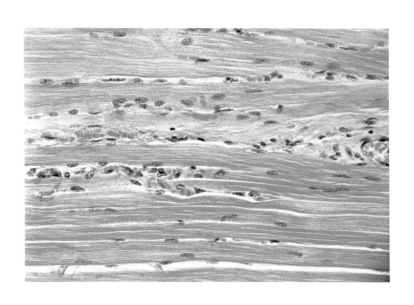

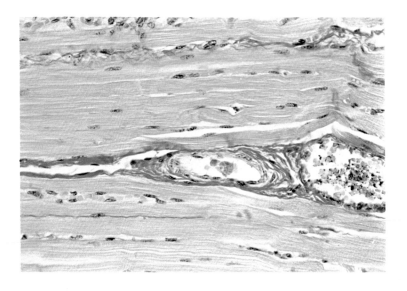

CONNECTIVE TISSUE

Collagen

22. Masson's trichrome (green or blue) (page 143)
× 100

Collagen: green (or blue)
Nuclei: blue-black
Cytoplasm: red
Muscle: red
Mucin: green or blue

This technique is a reasonable alternative to van Gieson and may be preferred. The darker staining makes single fibres easier to see.

23. MSB trichrome (page 136)
× 100

Nuclei: black
Red blood cells (erythrocytes): yellow
Fibrin: red
Muscle: red
Connective tissue: blue

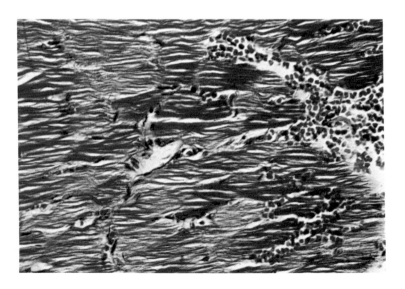

CONNECTIVE TISSUE

Elastic

24. Ehrlich's haematoxylin and eosin (page 122)
× 100

Elastic and other connective tissue fibres: pink to red
Nuclei: blue

25. Aldehyde fuchsin (page 112)
× 100

Elastic fibres: deep purple

This technique stains the elastic fibres deeply; even the finest are easy to observe. It has the added advantage that differentiation is not required.

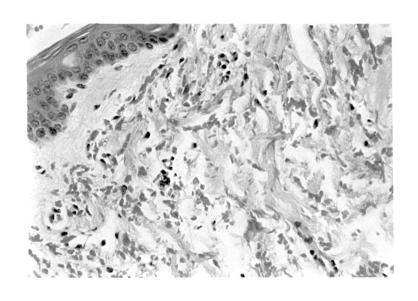

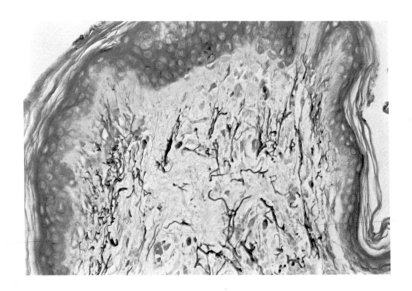

26. Orcein for elastic (page 150)
× 100

Elastic fibres: dark brown
Other tissues: as counterstained

This technique has the advantage of delicately and precisely staining elastic fibre but the finer fibres may be difficult to observe.

27. Verhoeff's haematoxylin for elastic and van Gieson's picro fuchsin (page 168)
× 100

Elastic fibres: black
Nuclei: blackish
Collagen (or as counterstained): red

The coarse fibres are well demonstrated but the finer ones are frequently lost during differentiation. Probably the only advantage of this method is that collagen fibres may be demonstrated simultaneously.

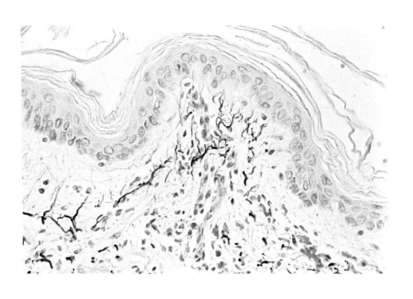

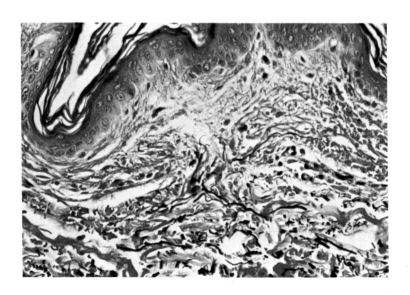

CONNECTIVE TISSUE

Muscle: striated

28. Ehrlich's haematoxylin and eosin (page 122)
× 100

Muscle and other connective tissue: pink to red
Nuclei: blue

29. Mallory's phosphotungstic acid haematoxylin (PTAH)
(page 138)
× 100

Muscle: dark blue

No differentiation is required and the striations are usually
easily demonstrated.

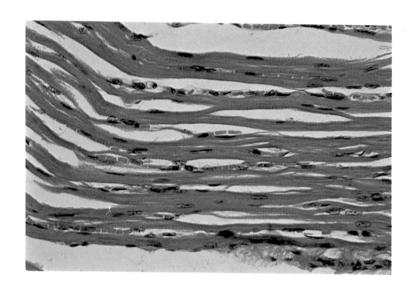

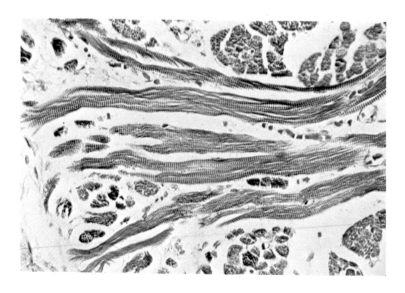

43

CONNECTIVE TISSUE

Muscle: striated

30. Masson's trichrome (blue or green) (page 143)
× 100

Muscle: red
Collagen: blue or green

Both collagen and muscle fibres are readily demonstrated by this technique.

31. Heidenhain's iron haematoxylin (page 134)
× 100

Muscle striations: blue-black

This illustration is included for the sake of completeness but the technique is not recommended as the staining is almost invariably patchy and for diagnostic purposes unreliable.

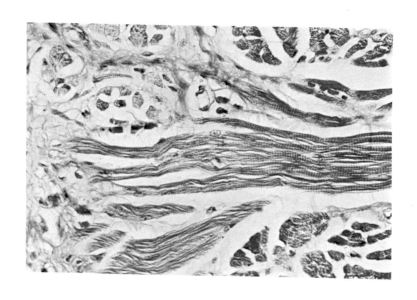

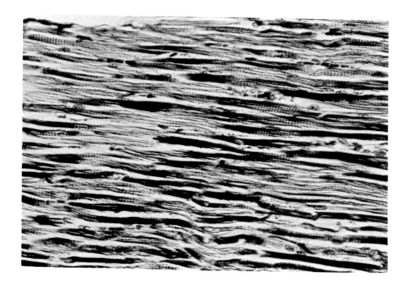

CONNECTIVE TISSUE

Muscle: smooth

32. Ehrlich's haematoxylin and eosin (page 122)
× 100

Muscle and other connective tissue fibres: pink to red
Nuclei: blue

33. Masson's trichrome (blue or green) (page 143)
× 100

Muscle: red
Collagen: blue or green
Nuclei: blue-black

Muscle fibres are differently stained from other connective
tissue fibres. Diagnostically this is an advantage when
searching for the odd muscle fibre.

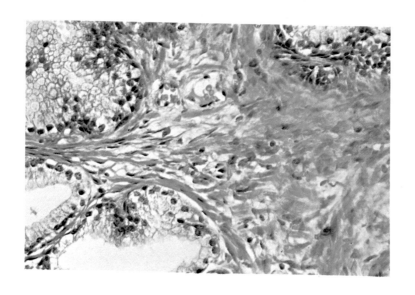

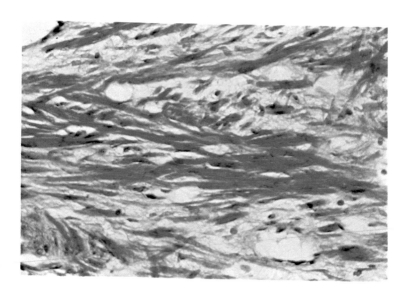

CONNECTIVE TISSUE

Oxytalan fibres

34. Oxytalan fibre, Fullmer's method (page 151)
× 100

Oxytalan fibres: brown

35. Oxytalan fibre, Fullmer's method (page 151)
× 160

Oxytalan fibres: brown

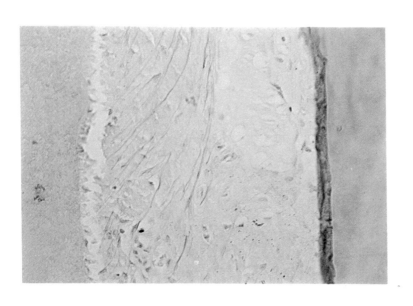

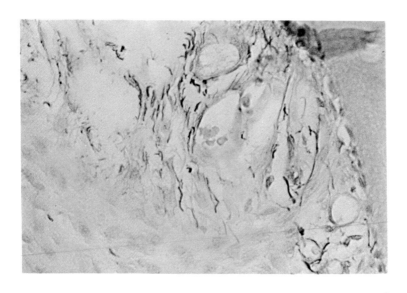

49

CONNECTIVE TISSUE

Reticulin

36. Foot's reticulin impregnation (Robb-Smith modification)
(page 125)
× 100

Reticulin: black
Collagen: golden brown

The sections are impregnated floating on the surface of the
solutions, thereby avoiding contamination of the section
by unwanted silver debris. The reticulin is clearly seen
against the untoned background.

37. Gomori's reticulin impregnation (page 127)
× 100

Reticulin: black
Nuclei: greyish
Collagen: dark grey to purple

Sections are impregnated while attached to slides in the
normal way. The technique is faster than Foot's
impregnation.

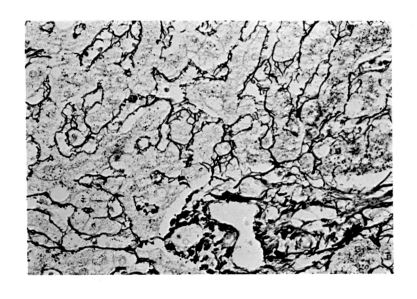

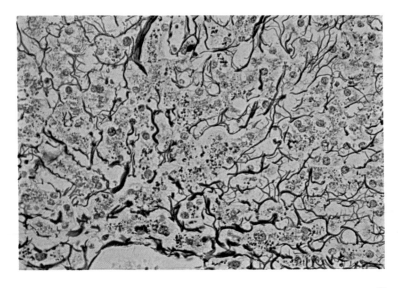

CARBOHYDRATES

Glycogen

38. Ehrlich's haematoxylin and eosin (page 122)
× 100

39. Best's carmine (page 116)
× 100

Glycogen: red
Nuclei: blue

40. Periodic acid Schiff (PAS) (page 152)
× 100

Positive materials including glycogen: reddish-purple
Nuclei: faint grey

Glycogen is only one of the many substances demonstrated by PAS. The results must always be checked using a control slide.

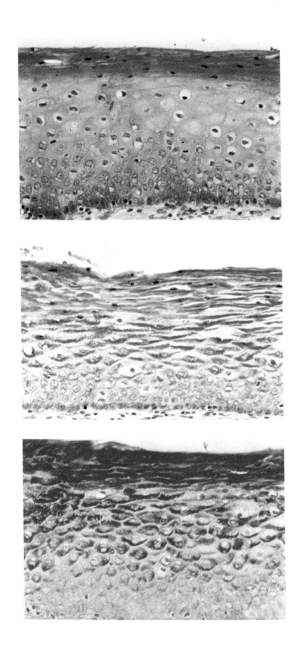

CARBOHYDRATES

Mucin

**41. Ehrlich's
haematoxylin and eosin**
(page 122)
× 100

**42. Southgate's
mucicarmine** (page 162)
× 100

Mucin: red
Nuclei: blue

This is the classic
method for demon-
strating epithelial
mucin but it is
unsuitable for con-
nective tissue mucin.

**43. Steedman's Alcian
blue** (page 163)
× 100

Mucin, mast cell
granules, cartilage
ground substance: blue
to blue-green

This and the following
techniques demonstrate
both epithelial and con-
nective tissue mucin.

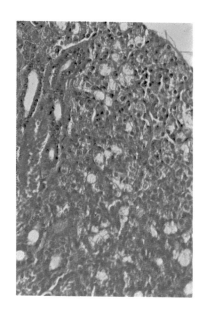

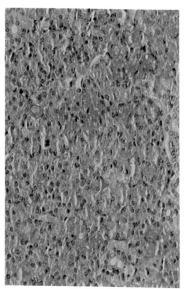

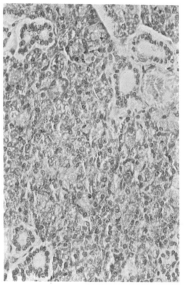

CARBOHYDRATES

Mucin

44. Hale's colloidal iron
(page 132)
$\times$ 100

Acid mucopoly-
saccharides: blue
Nuclei: red
Other structures:
shades of pink
Ferric iron, if present:
blue

**45. Periodic acid Schiff
(PAS)** (page 152)
$\times$ 100

Positive material:
reddish-purple
Nuclei: blue

The morphology will
confirm the presence of
epithelial mucin but
connective tissue mucin
will require confirma-
tion by the use of
another technique.

**46. Hughesdon's meta-
chromatic method**
(page 135)
$\times$ 100

Mucin (acid mucopoly-
saccharides):
metachromatically
Mast cell granules:
stained red-violet
Other tissue: blues

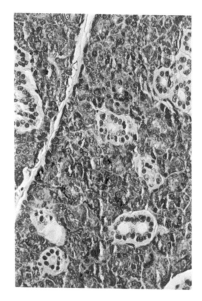

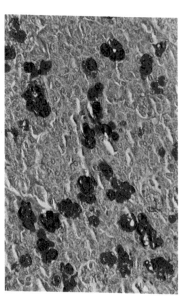

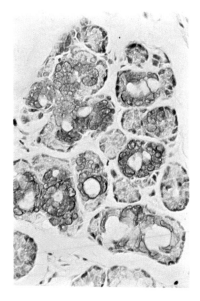

EXTRACELLULAR SUBSTANCES

Amyloid

**47. Ehrlich's
haematoxylin and eosin**
(page 122)
× 25

**48. Bennhold's Congo
red** (page 115)
× 25

Amyloid: dark pink-
red
Nuclei: blue

**49. 1 per cent methyl
violet metachromatic
staining** (page 146)
× 25

Amyloid: reddish-
purple
Other tissues: bluish

A simple flooding of
the section with stain
for 2 to 5 minutes and
rinse off allows
immediate confir-
mation of the presence
of amyloid.

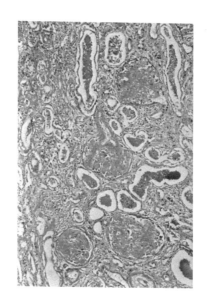

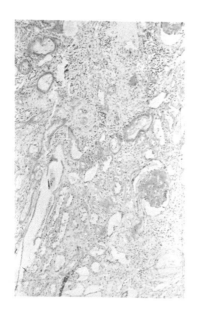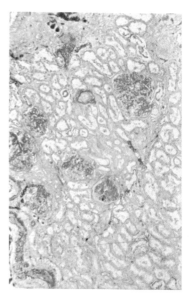

EXTRACELLULAR SUBSTANCES

Fibrin

**50. Ehrlich's
haematoxylin and eosin**
(page 122)
× 100

**51. Phosphotungstic
acid haematoxylin
(PTAH)** (page 138)
× 100

Fibrin: light to dark
blue

**52. MSB for fibrin
(Lendrum)** (page 136)
× 100

Nuclei: black
Erythrocytes: yellow
Fibrin: red
Fibrin 'old': bluish
Connective tissue: blue

This technique is
selective for fibrin but
not specific and the age
of the fibrin determines
its colour, veering from
the bright red of young
fibrin to the blue of old.

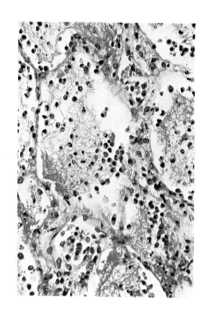

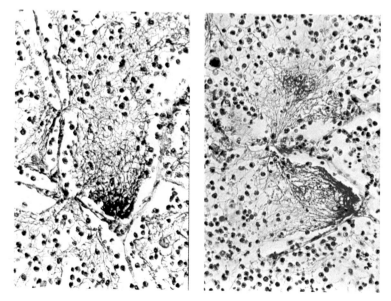

EXTRACELLULAR SUBSTANCES

Keratin

53. Ehrlich's haematoxylin and eosin (page 122)
× 25

54. Phloxine tartrazine (page 155)
× 25

Keratin: red
Nuclei: grey-blue

Keratin being phloxinophilic retains the stain tenuously; so do some other substances. The findings are therefore confirmed by morphology.

55. Rhodamine B for keratin (page 156)
× 25

Keratin: bright red
Nuclei and all basophilic material: blue or blue-black

This technique is suitable for fluorescence microscopy.

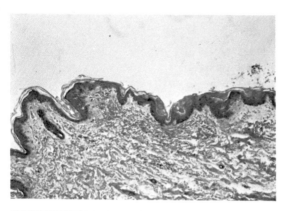

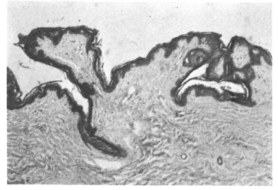

56. Feulgen reaction for DNA (page 124)
× 100

Deoxyribonucleic acid: magenta
Background: yellow

57. Feulgen reaction for DNA (page 124)
× 100

Control section without hydrolysis

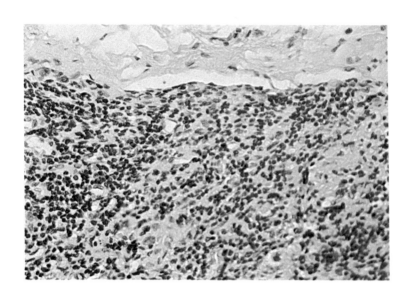

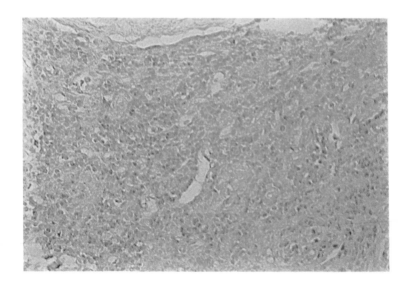

INTRACELLULAR

Eosinophils

**58. Ehrlich's
haematoxylin and eosin**
(page 122)
× 100

59. Azoeosin (page 114)
× 100

Eosinophil granules:
red
Nuclei: blue

**60. Differentiated eosin
for eosinophils** (page 124)
× 100

Eosinophil granules:
red
Nuclei: blue

The eosinophil retains
eosin long after it has
been differentiated
from other tissue.

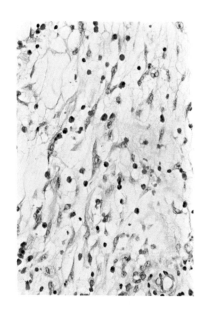

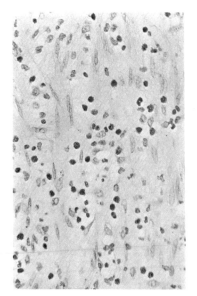
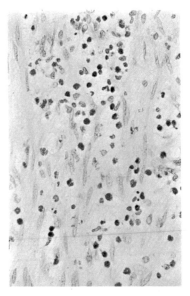

INTRACELLULAR

Mast cells

61. Ehrlich's haematoxylin and eosin
(page 122)
× 100

62. Hughesdon meta-chromatic method
(page 135)
× 100

Mast cell granules: red to violet
Some other substances also metachromatically staining
Other tissues: blue

63. Aldehyde fuchsin
(page 112)
× 100

Mast cell granules: deep purple

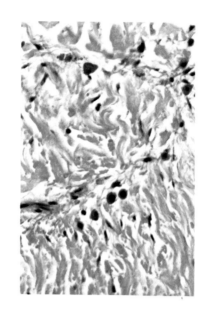

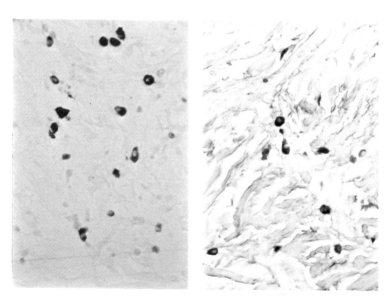

Mitotic figures

64. Ehrlich's haematoxylin and eosin (page 122)
× 100

65. Heidenhain's iron haematoxylin (page 134)
× 160

Nuclei and mitotic figures: blue-black

The haematoxylin is differentiated until the mitotic figures are clearly seen.

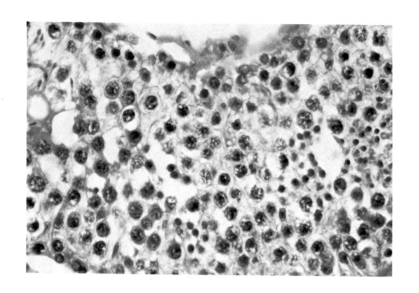

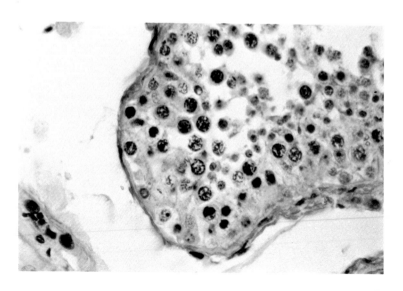

Mitotic figures

66. Heidenhain's iron haematoxylin (page 134)
× 160

Mitotic figures: blue-black

Onion root tip. This material is often used to show all the stages of mitosis.

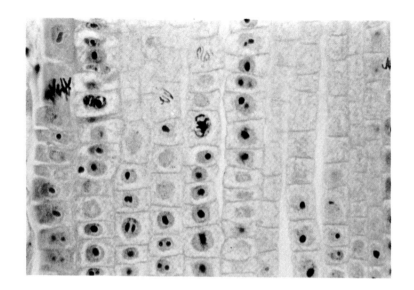

INTRACELLULAR

Paneth cells

67. Ehrlich's haematoxylin and eosin (page 122)
× 100

68. Phloxine tartrazine (page 155)
× 100

Paneth cell granules ⎫
Erythrocytes ⎪
Inclusion bodies ⎪
Russell bodies ⎬ red
Fibrin ⎪
Keratin ⎪
Pancreatic beta cells ⎭
Nuclei: grey-blue
Other tissues: yellow

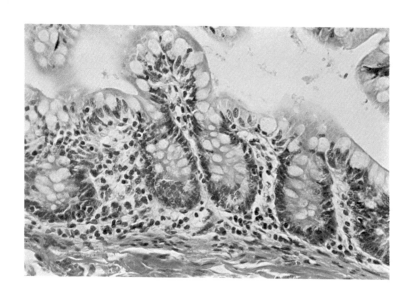

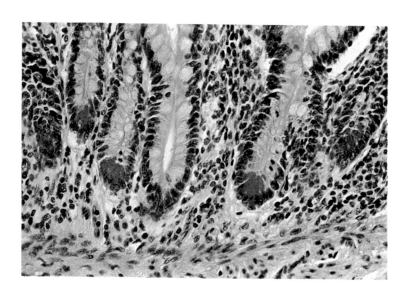

INTRACELLULAR

RNA (DNA) (plasma cells)

69. Ehrlich's haematoxylin and eosin (page 122)
× 160

70. Unna–Pappenheim (page 166)
× 160

RNA: red to reddish-purple
DNA: green to bluish-green

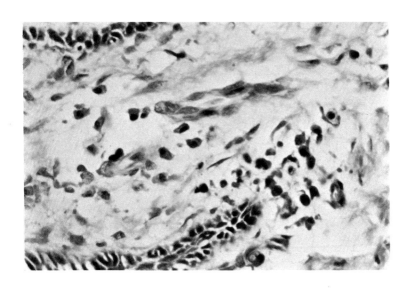

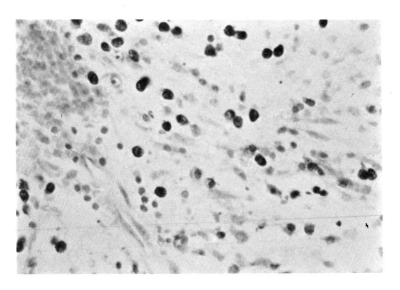

LIPIDS

**71. Ehrlich's
haematoxylin and eosin**
(page 122)
× 25

72. Sudan III and IV
(page 164)
× 25

Lipids: red
Nuclei: blue

73. Sudan black
(page 165)
× 25

Lipids: black

74. Nile blue sulphate
(page 148)
× 25

Neutral lipids: pink to
red
Acidic lipids: blue

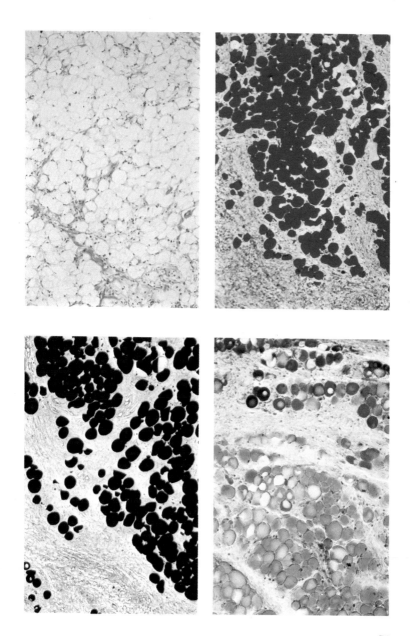

MICRO-ORGANISMS

Bacteria

**75. Ehrlich's
haematoxylin and eosin**
(page 122)
× 100

76. Gram's stain
(page 128)
× 100

Gram-positive
organisms: blue-black
Gram-negative
organisms: red

77. Gram Twort
(page 129)
× 100

Gram-positive
bacteria: dark blue
Gram-negative
bacteria: pink
Background: green

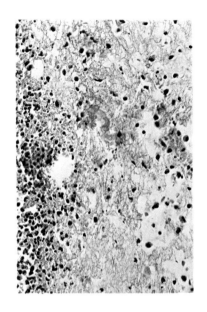

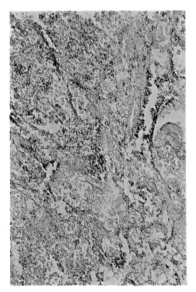

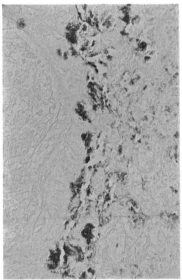

78. Ziehl Neelsen (page 172)
× 160

Acid alcohol-fast bacilli: red
Background: bluish
Erythrocytes: pale pink

79. Wade–Fite (page 171)
× 160

Acid-fast bacilli: red
Nuclei: blue

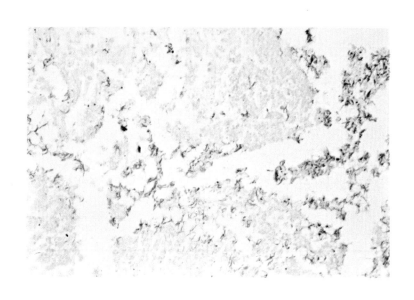

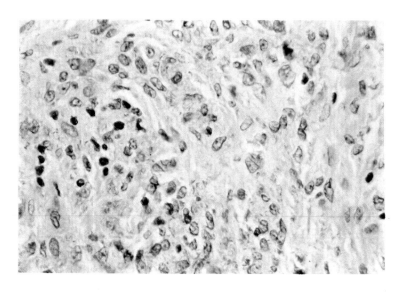

83

MICRO-ORGANISMS

Fungi

**80. Ehrlich's
haematoxylin and eosin**
(page 122)
× 100

**81. Silver impregnation
(Grocott)** (page 160)
× 100

Fungi walls outlined
black

82. Periodic acid Schiff
(page 152)
× 100

Fungi: pink

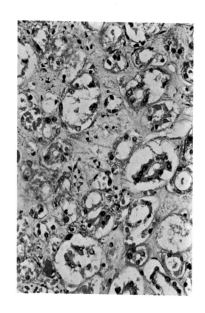

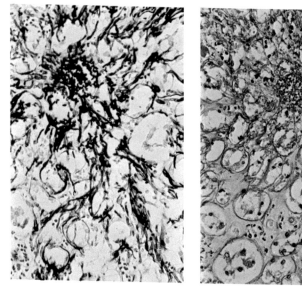

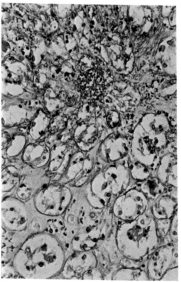

85

83. Ehrlich's haematoxylin and eosin (page 122)
× 160

84. Phloxine tartrazine (page 155)
× 160

Inclusion bodies
Erythrocytes
Paneth cells
Russell bodies   } red
Fibrin
Keratin
Pancreatic beta cells
Nuclei: grey-blue
Other tissue: yellow

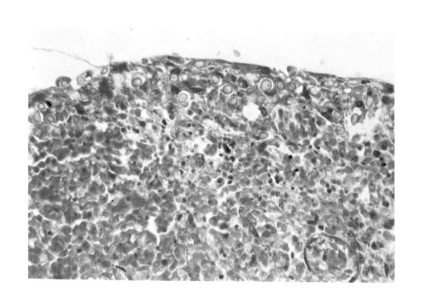

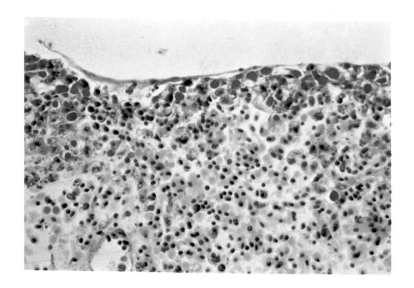

85. Marsland, Glees and Erikson's method for axons
(page 140)
× 100

Axons and dendrites: black
Other tissues { (if toned): light grey
(untoned): yellow-brown

86. Marsland, Glees and Erikson's method for axons
(page 140)
× 100

Axons and dendrites: black
Other tissues { (if toned): light grey
(untoned): yellow-brown

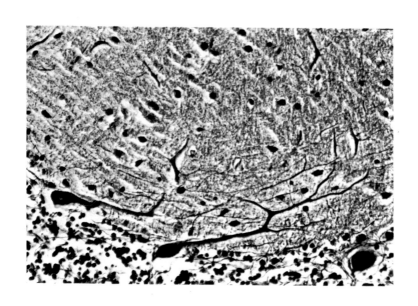

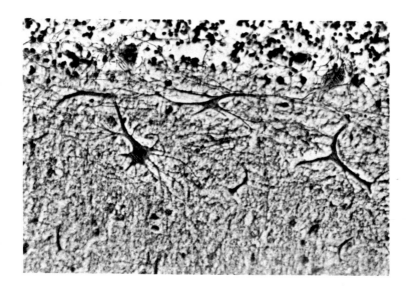

87. Weil's haematoxylin
(page 147)
× 100

Myelin sheaths: blue-black
Nuclei: greyish

Transverse section

88. Weil's haematoxylin
(page 147)
× 100

Myelin sheaths: blue-black
Nuclei: greyish

Longitudinal section

89. Sudan black
(page 165)
× 200

Myelin: black
Lipids: black

Transverse section

90. Sudan black
(page 165)
× 200

Myelin: black
Lipids: black

Longitudinal section

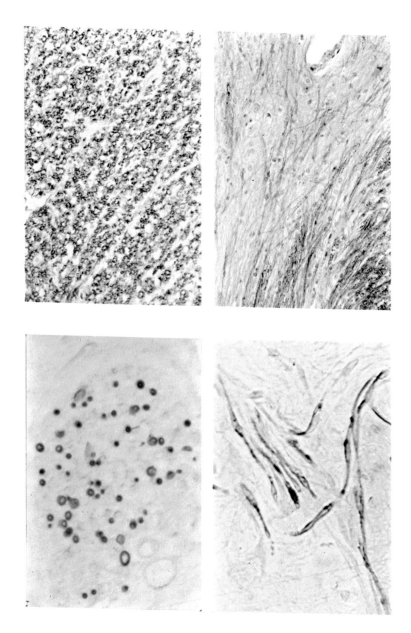

91

Nerve fibres

91. Bielschowsky, Smith and Godfrey modification
(page 117) × 100

Nerve fibres: black
Other tissues: yellow to light brown

92. Bielschowsky, Smith and Godfrey modification
(page 117) × 250

Nerve fibres: black
Other tissues: yellow to light brown

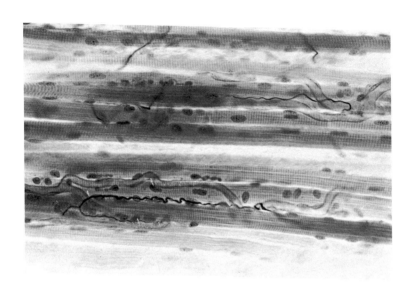

NERVE TISSUE

Nissl bodies and neuro-connective tissue

93. Nissl bodies (page 149)
× 100

Nissl bodies: light blue
Nuclei: light blue
Cytoplasm and background: colourless

94. Mallory's phosphotungstic acid, haematoxylin
(page 138)
× 100

Neuro-connective tissue: blue

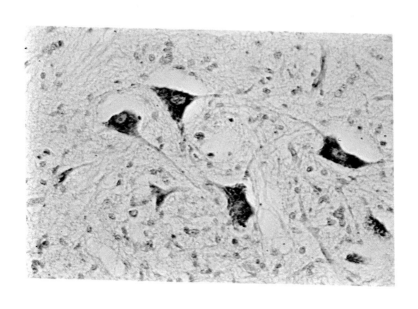

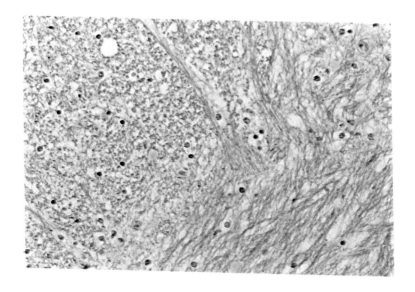

PIGMENTS

Bile

95. Ehrlich's haematoxylin and eosin (page 122)
× 100

96. Fouchet's method (page 120)
× 100

Bile pigment: olive green
Collagen: red
Other tissue: yellow

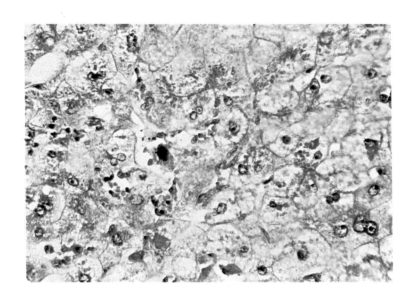

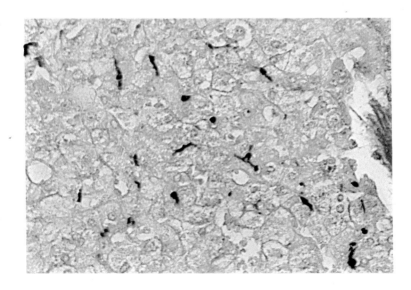

97

PIGMENTS

Copper

97. Ehrlich's haematoxylin and eosin (page 122)
× 100

98. Mallory and Parker's haematoxylin (page 138)
× 100

Copper: blue
Lead: dark blue grey

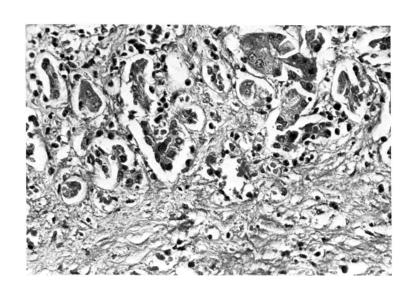

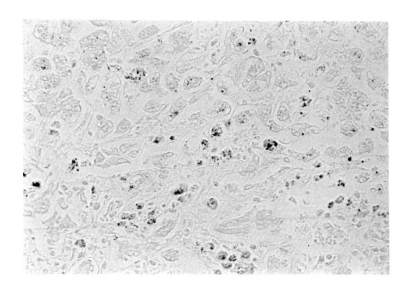

99. Ehrlich's haematoxylin and eosin (page 122)
× 100

100. Dunn–Thompson method (page 131)
× 100

Haemoglobin: greenish-blue-black

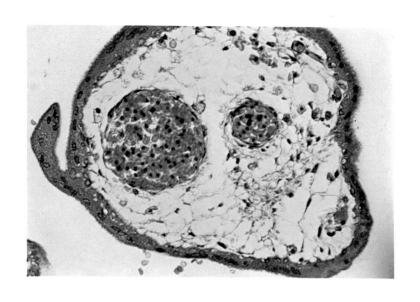

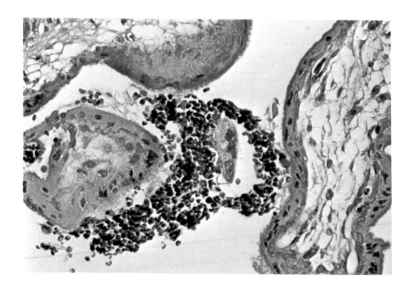

PIGMENTS

Free iron

---

101. Ehrlich's haematoxylin and eosin (page 122)
× 100

---

102. Perls' Prussian blue reaction (page 154)
× 100

Haemosiderin (ferric salt) ⎫
            (free iron) ⎬ dark blue
Other tissue: shades of pink
Other pigments: unstained

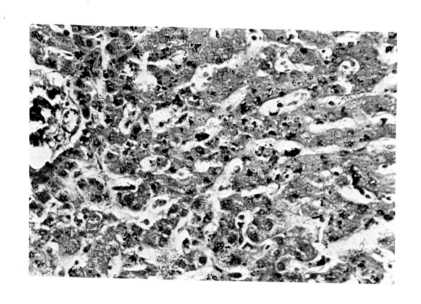

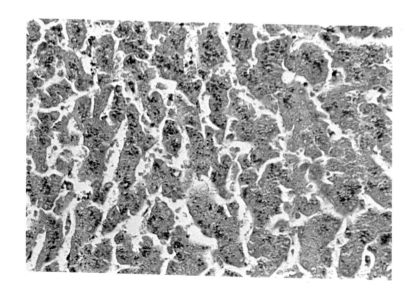

103

PIGMENTS

Melanin

**103. Ehrlich's
haematoxylin and eosin**
(page 122)
× 100

104. Masson–Fontana
(page 142)
× 100

Melanin (+ argentaffin
pigment): dark brown
to black
Nuclei: red
Other tissue: pinkish

This technique is used
to demonstrate the
produced pigment, not
the cell producing it.

**105. Dopa-oxidase
reaction**
× 100

Melanin-forming cells
(+ non-pigmented
melanomas): dark
brown to black

This illustration is
included to emphasise
the fact that the
demonstration of
melanin does not show
the melanocytes or
non-pigmented
melanin.
   The technique is
specialised and outside
the scope of this book;
details are therefore not
included.
(Ref. BECKER, S.W.,
PRAVER, L.L. and
THATCHER, H. 1935,
Archs. Derm. Syph. **31**,
190.)

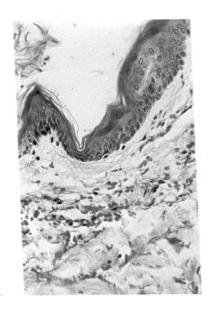

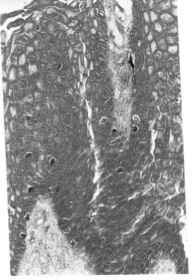

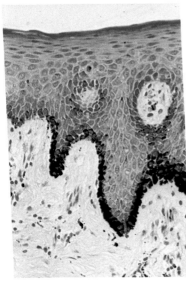

PART 2

Staining and impregnation technique: working instructions

BASIC REQUIREMENTS

Adequate weak alkaline, running cold water supply, or tap water substitute

Running hot water

Absolute alcohol in bulk

Xylene in bulk

Solvents as indicated in staining technique

Dry dye powders

Mounting media

Chemicals as indicated by staining techniques to be used

Hotplate (gas or electric)

Mounting needle

Conical flasks

Forceps

Cold staining rack

Coverslips of various sizes

Stock bottles of stain solutions

Balance

Measuring cylinders, graduated

Pestle and mortar

Beakers, lipped, various

Funnels, various

Test-tubes

Filter paper

Sand, washed

Slide trays

Glass dishes for bulk staining, dehydrating and clearing

Staining racks, stainless steel

Gauze or fluff-free cloth

Glass petri dishes

Watch glasses

Pasteur pipettes

Graduated pipettes

Teats

Writing diamond

Grease pencil

Labels
Test-tube rack
Bunsen burner
Coplin jars
Plastic squeeze bottles
Water vacuum pump
Microscope, staining
Microscope (ultra-violet, birefringence and photomicro-
graphy desirable)
Oil, immersion
Refrigerator, small
Incubator, small – ranging from 20 °C to 60 °C
Tripod, bunsen
Gauze, asbestos wire

REPETITIVE TERMINOLOGY

When staining techniques are presented in tabulated style, it is
acknowledged practice to use abbreviations instead of repetitive
detail to indicate procedures common to most staining techniques.

1. Sections to water (or *y* per cent alcohol)
Almost invariably sections are taken into water or alcohol of the
appropriate strength prior to treatment with the stain. Many staining
solutions are aqueous while others have a high alcohol content. To
take sections from water into an alcoholic staining solution is to risk
precipitation of the dye on to the section or otherwise to affect the
staining properties of the solution.
   (a) Warm sections to soften the paraffin wax, but do not melt.
   (b) Remove paraffin wax with xylene.
   (c) Remove xylene with absolute alcohol followed by descending
        grades of alcohol to water/alcohol as indicated in text.

2. Dehydrate, clear and mount
On completion of staining, the section is usually permanently sealed
under a thin glass coverslip, the mountant being a natural resin or a
synthetic one, almost invariably not miscible with water or alcohol.

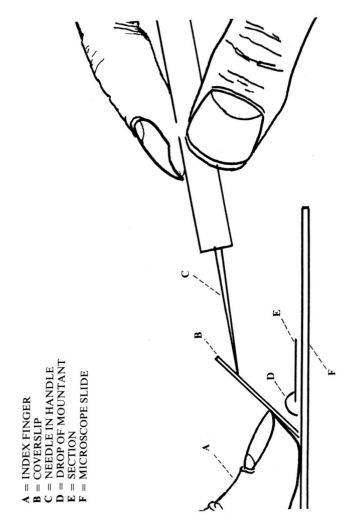

A = INDEX FINGER
B = COVERSLIP
C = NEEDLE IN HANDLE
D = DROP OF MOUNTANT
E = SECTION
F = MICROSCOPE SLIDE

Fig. 106.

110

As many of the completed stains are affected by prolonged contact with water or alcohol, it is essential that the mountant contains no trace of these.

(a) *Dehydrate*. Sections are taken through ascending grades of alcohol to absolute alcohol, to ensure complete removal of water.

(b) *Clear*. Treat sections with two changes of xylene to remove all trace of alcohol.

(c) *Mount*. The surplus xylene is drained from the section, care being taken not to allow it to dry. The slide is placed, section uppermost, on the bench, a drop of mountant being placed on the section towards one end. The coverslip is then held at an angle to the slide with one edge resting on the slide clear of the section. With the aid of a needle the coverslip is carefully lowered, its weight causing the mountant to flow across the section without trapping bubbles (Fig. 106). Endeavour not to move the coverslip once it is in contact with the section to avoid damage to the section. Whether the section is gently heated to assist evaporation of the solvent or left to dry naturally will depend on the mountant or requirement of the laboratory.

3. Dip, rinse or wash

(a) *Dip*. Section in and out of solution without delay.

(b) *Rinse*. Section in and out of solution with short delay.

(c) *Wash*. Section in and out of solution, keeping immersed for several minutes each time, unless otherwise indicated.

4. Preparation of staining solution

Unless otherwise stated, distilled water is always used for preparing the solutions.

ALDEHYDE FUCHSIN FOR ELASTIC FIBRES

Solutions required

(a) 2.5 per cent sodium thiosulphate

(b) Lugol's iodine

Iodine	1 gm
Potassium iodide	2 gm
Distilled water	100 ml

(c) Aldehyde fuchsin

Basic fuchsin	0.5 gm
70 per cent alcohol	100 ml
Conc. hydrochloric acid	1 ml
Paraldehyde	1 ml

Dissolve the basic fuchsin in the alcohol, add the hydrochloric acid and paraldehyde. Keep at 20 °C for 1–3 days until the solution assumes the deep purple colour indicating that it is ready for use. Keeps 2–3 months at 4 °C.

Technique

1 Sections to water.
2 Lugol's iodine   10 minutes.
3 Wash in tap water.
4 Decolorise in sodium thiosulphate.
5 Wash in tap water.
6 Transfer to 70 per cent alcohol.
7 Stain in Coplin jar of aldehyde fuchsin   10 minutes.
8 Wash in several changes 95 per cent alcohol.
9 Counterstain if desired.
10 Dehydrate, clear and mount.

Results

Elastic fibres: deep purple

Mast cell granules and some mucopolysaccharides: deep purple

Other tissues according to counterstain.

Notes
(a) The basic fuchsin must be pure.
(b) Used in this way the elastic fibres are demonstrated without the need for differentiation.

Reference
GOMORI, G. 1950, *Amer. J. clin. Path.* **20**, 665.

ALIZARIN RED S

Solutions required
2 per cent alizarin red S in distilled water adjusted to pH 4.1–4.3 with dilute ammonium hydroxide.

Technique
1 Section to water.
2 Rinse in distilled water.
3 Stain with alizarin red S solution   30 seconds to 5 minutes. Examine under the microscope frequently.
4 When orange-red staining is strong but not diffuse, blot with filter paper.
5 Dehydrate with acetone 20 seconds, then equal parts acetone and xylol 20 seconds, followed by final clearing in xylene.
6 Mount.

Results
Sites of calcium: orange-red

Reference
McGEE-RUSSELL, S.M. 1958, *J. Histochem. Cytochem.* **6**, 22.

AZOEOSIN

Solutions required
(a) Celestine blue
(b) Mayer's haemalum
(c) Azoeosin

      Azoeosin   1 gm
      Distilled water   200 ml

Mix azoeosin and distilled water and bring to boiling point. Then add the following:

      Methyl alcohol   16 ml
      Glycerine   4 ml

This solution keeps well.

Technique
1 Sections to water.
2 Stain Celestine blue, Mayer's haemalum sequence. (See page 167 stages 1 to 5.)
3 Wash in water.
4 Transfer to absolute alcohol.
5 Stain in azoeosin   30 minutes.
6 Rinse and dehydrate in absolute alcohol.
7 Clear in xylol and mount.

Results
Eosinophils: red
Nuclei: blue

MODIFIED BENNHOLD CONGO RED

Solutions required
(a) Mayer's haemalum
(b) Alkaline solution
   Saturated solution of sodium chloride in 80 per cent alcohol. For use take 50 ml of stock solution and add 0.5 ml of 1 per cent sodium hydroxide. Filter and use immediately.
(c) Congo red solution
   Saturated solution of Congo red and sodium chloride in 80 per cent alcohol. For use add 0.5 ml of 1 per cent sodium hydroxide to 50 ml of stock staining solution. Filter and use immediately.

Technique
1 Sections to water.
2 Stain lightly in Mayer's haemalum.
3 Wash in water.
4 Treat with solution (b)   20 minutes.
5 Stain in Congo red mixture   20 minutes.
6 Dehydrate rapidly in 3 changes of absolute alcohol.
7 Clear and mount.

Results
Amyloid: dark pink–red
Nuclei: blue
Elastic fibres and some granules may be stained pink–red

Notes
The stock solutions (b) and (c) may be stored and must be at least 24 hours old before use. The working solution must be freshly prepared.

References
BENNHOLD, H. 1922, *Munch. med. Wschr.* **69**, 1537.
PUCHTLER, *et al.* 1962, *J. Histochem. Cytochem.* **10**, 355.

BEST'S CARMINE METHOD

Solutions required
(a) Ehrlich's haematoxylin
(b) 1 per cent acid alcohol
(c) Best's carmine (stock solution)

Carmine	6 gm
Potassium carbonate	3 gm
Potassium chloride	15 gm
Distilled water	180 ml

Mix together and boil in a large flask for 5 minutes.

Ammonia (0.880 s.g.)	60 ml

Cool, add the ammonia.
This stock solution keeps in the refrigerator for some months.
Before use always check, using a control slide.
(d) Best's carmine (working solution)

Stock solution	2 parts
Ammonia (0.880 s.g.)	2 parts
Absolute methyl alcohol	3 parts

(e) Best's differentiator

Absolute methyl alcohol	40 ml
Absolute ethyl alcohol	80 ml
Distilled water	100 ml

Technique
1 Sections to water.
2 Stain in Ehrlich's haematoxylin   20 minutes.
3 Differentiate in 1 per cent acid alcohol.
4 Wash in water and check staining, using microscope.
5 Stain in Best's carmine working solution in a
  closed Coplin jar   5–10 minutes.
6 Differentiate in Best's differentiator until background is clear   1–5 minutes.
7 Without touching water, transfer section to absolute alcohol.
8 Complete dehydration, clear and mount.

Results
Glycogen: red
Nuclei: blue

Notes
Conflicting opinions are held regarding the usefulness of this technique, and whether or not special precautions are required in the fixation and sectioning of this material. However, if the results are checked using similar control material the result using the technique described is valid and useful.

Reference
BEST, F. 1906, *Z. Wise M. Krusk,* **23**, 319.

BIELSCHOWSKY, SMITH AND GODFREY MODIFICATION

Solutions required
(a) Trichloracetic acid, formaldehyde solution

Trichloracetic acid	5 gm
Formaldehyde	25 ml
Water	75 ml

(b) 20 per cent silver nitrate
(c) 10 per cent formaldehyde (1 part formaldehyde to 3 parts water)
(d) 5 per cent formaldehyde
(e) 2 per cent sodium thiosulphate
(f) Ammoniacal silverbath
    To 5 ml of 20 per cent silver nitrate add 0.5 ml of ammonia (0.88 s.g.) drop by drop from a burette, shaking the silver solution all the time. Then continue to add ammonia as before until the precipitate first formed just redissolves. Seven more drops of ammonia are then added and the solution filtered.
(g) Fixative
    10 per cent formaldehyde saturated with magnesium carbonate, filter just before use.

Technique

1 Fix fresh tissue in 'neutral' 10 per cent formaldehyde (2 days minimum).

2 Decalcify in 5 per cent trichloracetic acid in 10 per cent formaldehyde if calcified tissue is to be sectioned.

3 Cut frozen sections 10 to 50 microns thick. The sections collected into and stored in 10 per cent 'neutral' formaldehyde until required.

4 Suppression of collagen staining and more clearly defined nerve fibres result from treatment of sections in the decalcifying solution for 60 minutes at 60 °C. Do not treat more sections than can be handled within a reasonable time.

5 Take 2 sections at a time from the acid formaldehyde bath and rinse in distilled water for a maximum of   10 seconds.

6 Transfer sections to 5 ml of filtered 20 per cent silver nitrate at 37 °C for   10 minutes.
A fresh solution should be used after four to six sections.

From stage 7 to 13 all processing is carried out at 14 °C by standing the containers of solutions in a controlled waterbath. All solutions should be kept at this temperature ready for use.

7 The sections are passed for a total of $1\frac{1}{2}$ minutes without washing through three baths, each containing 40 ml of 10 per cent formaldehyde. In the first and second baths a white precipitate forms and, while constantly moving the section, care must be taken to avoid passing the section through the precipitate.
Fresh formaldehyde baths must be used in stages 7 and 11.

   (a) 10 per cent formaldehyde   5–10 seconds.
   (b) 10 per cent formaldehyde   20–30 seconds.
   (c) 10 per cent formaldehyde to a total of $1\frac{1}{2}$ minutes.

8 5 per cent formaldehyde   30 seconds.

9 Rinse in distilled water   5–10 seconds.

10 Treat sections in fresh ammoniacal silverbath   2 minutes.

11 Without washing, the sections are passed through four baths of 5 per cent formaldehyde, 40 ml in each. Follow the instructions for stage 7.

   (a) 5–10 seconds.

(b) 30 seconds.

(c) 30 seconds.

(d) 10 minutes.

12 Wash section in distilled water. (Sections may remain in distilled water while other sections are being processed to this stage.)

13 2 per cent thiosulphate   2 minutes.

14 Wash in distilled water.

15 Dehydrate.

16 Clear in carbol xylene.

17 Mount in colophonium in turpentine.

Results

Nerve fibres: black

Collagen: golden-yellow

Muscle: red-brown

Nuclei: brown

Notes

(a) The formaldehyde solution used in this technique is a 40 per cent commercial solution, dilutions being made from this percentage; i.e. a 10 per cent solution is 1 part formaldehyde to 3 parts water. The fixative is kept saturated with magnesium carbonate and filtered before use.

(b) Owing to the thickness of the sections and the ramifications of the nerve fibres it is an advantage to mount the section between two coverslips so that both sides of the section may be examined with a high-power lens.

(c) Silver impregnations for nerve fibres are capricious and frequently fail. This technique has a high success rate.

References

BIELSCHOWSKY, M. 1904, *J. Psychol. Neurol. (Lpz.)* **3**, 169.

BIELSCHOWSKY, M. 1909, *J. Psychol. Neurol. (Lpz.)* **12**, 135.

SMITH, A., and GODFREY, J.L. 1963, *J. Med. Lab. Tech.* **20**, 291–4.

BILE, FOUCHET'S METHOD

Solutions required
(a) 25 per cent trichloracetic acid in distilled water   100 ml
     10 per cent ferric chloride in distilled water   10 ml
     Mix and filter just before use.
(b) van Gieson's picro fuchsin.

Technique
1 Section to water.
2 Ferric chloride–trichloracetic acid mixture   5 minutes.
3 Wash in distilled water   3 minutes.
4 Wash in running tap water   3 minutes.
5 Stain in van Gieson's picro fuchsin   5 minutes.
6 Dehydrate, clear and mount.

Results
Bile pigment: olive green
Collagen: red
Other tissues: yellow

Notes
(a) Paraffin sections may be used to demonstrate conjugated bile
     pigment but frozen sections must be used for demonstrating
     unconjugated.
(b) van Gieson's picro fuchsin used as a counterstain helps to
     prevent fading of the reaction.

References
FOUCHET, A. 1917, *Compt. rend. Soc. biol.* **80**, 826–8.
FOUCHET, A. 1918, *J. pharm. et chim. 7 ser.* **18**, 19–20.
HALL, M.J. 1960, *Am. J. clin. Path.* **34**, 313.

COLE'S HAEMATOXYLIN

Solution required

Haematoxylin	1.5 gm
Distilled water	250 ml
Add 1 per cent iodine in 95 per cent alcohol	50 ml
Dissolve with the aid of gentle heat.	
Saturated solution of aluminium ammonium sulphate	700 ml

Mix the two solutions together and bring to the boil. Cool immediately and filter.

Technique

Use technique as detailed for Ehrlich's haematoxylin. Cole's haematoxylin 10 minutes.

Notes

(a) Keeps well.

(b) May be used in place of Mayer's haemalum in the Celestine blue sequence.

Reference

COLE, E.C. 1943, *Stain Tech.* **18**, 125.

EHRLICH'S ACID HAEMATOXYLIN AND EOSIN

Solutions required

(a) Ehrlich's acid haematoxylin

Haematoxylin	14 gm
Absolute alcohol	700 ml

Dissolve with the aid of gentle heat.

Glycerine	700 ml
Distilled water	700 ml
Aluminium potassium sulphate to excess, approx.	150 gm

Dissolve with the aid of gentle heat.

Mix the two solutions together and add glacial acetic acid   10 ml

The solution will 'ripen' naturally if left for several weeks exposed to sunlight in a loosely stoppered bottle, but more commonly partial oxidation is achieved by the addition of 2.1 gm of sodium iodate; the solution may then be used within 24 hours

(b) 1 per cent eosin, yellowish, in distilled water

Technique

1 Section to water.
2 Stain in Ehrlich's acid haematoxylin   20–30 minutes.
3 Wash in running tap water until blue.
4 Differentiate in 1 per cent hydrochloric acid in 70 per cent alcohol   2–3 dips.
5 Wash in running tap water until blue.
6 Examine under the microscope (see note (b)).
7 Wash in running tap water for at least   10–15 minutes.
8 Counterstain in 1 per cent eosin   2–5 minutes.
9 Differentiate in running tap water until nearly desired colour is obtained.
10 Complete differentiation and commence dehydration in 95 per cent alcohol.
11 Dehydrate, clear and mount.

Results
Nuclei: blue-black
Cartilage: bluish
Incremental lines in bone: blue
Cytoplasm and connective tissue fibres: shades of pink and red

Notes
(a) Ehrlich's haematoxylin is used almost invariably regressively (overstaining) as is eosin, followed by differentiation to required degree.
(b) After differentiation in acid alcohol the sections are washed in tap water until they change from pink to blue. When differentiation is correct the sections are given a prolonged time in running tap water to ensure removal of the acid. If over-differentiated repeat stages 2–6; if under-differentiated repeat stages 4–6.
(c) The length of time sections are stained in haematoxylin and eosin will depend on the fixative used and whether or not the tissue required decalcification. Haematoxylin staining is inhibited by decalcification and eosin staining is enhanced.

Reference
EHRLICH, P. 1886, *Z. f. wiss. Mikr.* **3**, 150.

DIFFERENTIATED EOSIN FOR EOSINOPHILS

Solutions required
(a) Mayer's haemalum
(b) 1 per cent eosin, yellowish

Technique
1 Sections to water.
2 Lightly stain nuclei with Mayer's haemalum.
3 'Blue' in running water.
4 Stain in 1 per cent eosin   15 minutes.
5 Differentiate in water until only the eosinophils are stained.
6 Dehydrate, clear and mount.

Results
Eosinophils: red
Nuclei: pale blue

FEULGEN REACTION FOR
DEOXYRIBONUCLEIC ACID (DNA)

Solutions required
(a) Schiff's reagent
(b) Sulphurous acid
(c) N/1 hydrochloric acid
(d) Saturated tartrazine in cellusolve

Technique
1 Sections to water.
2 Rinse in N/1 hydrochloric acid.
3 Treat section in N/1 hydrochloric acid previously
   heated and maintained at 60 °C   5–10 minutes.
4 Rinse in distilled water.
5 Schiff's reagent $1–1\frac{1}{2}$ hours.
6 Wash in 3 changes of sulphurous acid, 2 minutes
   in each change.

7 Counterstain in saturated tartrazine in cellosolve
   (2–ethoxyethanol)   1 minute.
8 Rinse in water.
9 Dehydrate, clear and mount.

Results
DNA: magenta
Background: yellowish

Notes
(a) The time period for treatment of the section with heated N/1
    hydrochloric for critical work must be determined by controlled
    trials.
(b) A control section untreated with N/1 hydrochloric should be
    stained with Schiff's reagent.

Reference
FEULGEN, R., and ROSSENBECK, H. 1924, *Z. phys. chem.* **135**, 203.

FOOT'S RETICULIN IMPREGNATION TECHNIQUE: ROBB-SMITH'S MODIFICATION

Solutions required
(a) 10 per cent ammonia 0.88 s.g.
(b) 0.25 per cent potassium permanganate
(c) 1.5 per cent oxalic acid
(d) 10 per cent silver nitrate
(e) 30 per cent formalin, neutral (1–3 parts)
(f) Ammoniacal silver solution
        10 per cent silver nitrate   8 ml
        10 per cent sodium hydroxide   6 drops
    Add conc. 0.88 s.g. ammonia drop by drop until precipitate is just
    dissolved. Make up to 28 ml with distilled water and filter.

Technique

1 Float paraffin-embedded sections on to distilled water. (A Petri dish is used throughout the technique. Each fluid is sucked away by means of a Pasteur pipette attached to a water vacuum pump, care being taken not to suck up the sections.)
2 Ammonia 10 per cent   5 minutes.
3 Wash with three changes of distilled water.
4 Potassium permanganate 0.25 per cent   25 minutes.
5 Wash once with distilled water.
6 Bleach with oxalic acid 1.5 per cent until white.
7 Wash in four changes of distilled water.
8 Sensitise in silver nitrate 5 per cent   1 hour.
9 Wash in three changes of distilled water.
10 Impregnate in ammoniacal silver solution   25 minutes.
11 Wash in three changes of distilled water.
12 Reduce in formalin 30 per cent   3 minutes.
13 Wash in tap water   30 seconds.
14 Wash in distilled water.
15 Mount sections on slides, blot dry and heat sections on low heat hotplate to complete drying and adhesion.
16 Remove wax in xylene and mount.

Results

Reticulin: black
Collagen: golden brown

Notes

(a) This is an excellent method for the demonstration of reticulin, having the advantage that as the sections are floating on the surface of the solutions there is almost no risk of contamination of sections by 'falling' deposits.
(b) The colour contrast between reticulin and other tissues is excellent without the need for toning and counterstaining.
(c) Care must be taken when making up the ammoniacal silver. The solution must be well shaken between the addition of each drop of ammonia.

References

Foot, W.C. 1929, *J. techn. Meth.* **12**, 117.
Robb-Smith, A.H.T. 1937, *J. Path. Bact.* **45**, 312.

GOMORI'S RETICULIN IMPREGNATION TECHNIQUE
(Modification of Perdrau)

Solutions required

(a) 1 per cent potassium permanganate
(b) 2 per cent potassium metabisulphite
(c) 2 per cent ferric ammonium sulphate
(d) 10 per cent formalin, neutral
(e) 0.2 per cent gold chloride
(f) 2.5 per cent sodium thiosulphate
(g) 10 per cent potassium hydroxide
(h) Ammoniacal silver solution

> 10 per cent silver nitrate   10 ml
> 10 per cent sodium hydroxide   2 ml

Mix the silver nitrate and sodium hydroxide.
Add ammonia drop by drop until the precipitate first formed just dissolves, then add sodium hydroxide 10 per cent drop by drop until the new precipitate hardly dissolves with shaking. Add an equal volume of distilled water and filter.

Technique

 1 Sections to water.
 2 Oxidise in potassium permanganate 10 per cent   2 minutes.
 3 Rinse in water.
 4 Decolorise in potassium metabisulphite   1 minute.
 5 Prolonged wash in water.
 6 Sensitise in ferric ammonium sulphate 2 per cent   1 minute.
 7 Prolonged wash in tap water, followed by 2 changes of distilled water.
 8 Impregnate in ammoniacal silver solution   1 minute.
 9 Rinse in distilled water   5 seconds.
10 Reduce the silver in formalin 10 per cent by holding the slide at an acute angle and pouring the formalin down it; then resting the slide section uppermost on the cold staining rack, flood with formalin   3 minutes.
11 Wash in tap water.
12 Tone in gold chloride 0.2 per cent   5–10 minutes.

13 Rinse in distilled water.
14 Treat with potassium metabisulphite 1 minute.
15 Rinse in distilled water.
16 Treat with sodium thiosulphate 2.5 per cent   1–2 minutes.
17 Wash in water, dehydrate, clear and mount.

Results
Reticulin: black
Nuclei: greyish
Collagen: dark grey-purple

Note
After stage 9, it is advisable to hold the section at an acute angle when first applying a fresh solution so that any contaminating deposit that forms will not settle on the section.

References
GOMORI, G. 1937, *Amer. J. Path.* **13**, 993.
PERDRAU, J.R. 1921, *J. Path. Bact.* **14**, 117.

GRAM'S STAIN

Solutions required
(a) Lugol's iodine
    Iodine   1 gm
    Potassium iodide   2 gm
    Distilled water   100 ml
    Dissolve the potassium iodide in 5 ml of distilled water, dissolve the iodine in this, make up to 100 ml for Lugol's iodine or 300 ml for Gram's iodine.
(b) 1 per cent neutral red
(c) 0.5 per cent methyl violet

Technique
1 Section to water (plus known control).
2 Stain in 0.5 per cent methyl violet   1–3 minutes.

3 Rinse in water.
4 Lugol's iodine   1–3 minutes.
5 Differentiate in acetone, wash immediately in
   running water   1–2 seconds.
6 Counterstain in neutral red   1 minute.
7 Wash and differentiate in water.
8 Dehydrate clear and mount.

Results

Gram-positive organisms: blue-black
Gram-negative organisms: red

Notes

(a) Great care must be taken with the differentiation. Absolute
    alcohol may be used which is slower, up to half a minute.
(b) It is essential that a known control slide is simultaneously
    stained.

Reference

GRAM, C. 1884, *Forlschr. Med.* **2**, 185.

GRAM TWORT

Solutions required

(a) Modified Twort's stain
         0.2 per cent alcoholic neutral red   90 ml
         0.2 per cent alcoholic fast green   10 ml
    Mix the neutral red and fast green.
    For use dilute one part stain to 3 parts distilled water.
(b) 1 per cent aniline crystal violet
(c) Gram's iodine
(d) 2 per cent acetic acid

Technique

1 Sections to distilled water.	
2 Stain in 1 per cent aniline crystal violet	3–5 minutes.
3 Wash quickly in distilled water.	
4 Treat with Gram's iodine	3 minutes.
5 Wash quickly in distilled water and blot dry.	
6 Decolorise with 2 per cent acetic acid until no more colour floods from the section.	
7 Wash quickly in distilled water.	
8 Counterstain in diluted modified Twort's stain	5 minutes.
9 Differentiate with 2 per cent acetic acid until no more neutral red leaches from the section	15–30 seconds.
10 Dehydrate, clear and mount.	

Results

Gram-positive bacteria: dark blue
Gram-negative bacteria: pink
Cytoplasm: light green
Nuclei: red
Erythrocytes: green

Notes

(a) The differentiation is easier in this method.
(b) Filter aniline crystal violet solution before use.
(c) Use a known control section.

References

TWORT, F.W. 1924, *J. State Med.* **5**, 228.
OLLETT, W.S. 1947, *J. Path. Bact.* **59**, 357.
OLLETT, W.S. 1951, *J. Path. Bact.* **63**, 166.

HAEMOGLOBIN, DUNN-THOMPSON METHOD

Solutions required

(a) Ripened alum haematoxylin

Haematoxylin	0.25 gm
Aluminium potassium sulphate	5.00 gm
Distilled water	100 ml

(b) 4 per cent ferric ammonium sulphate in distilled water

(c) van Gieson's picro fuchsin

Technique

1	Section to water.	
2	Stain in alum haematoxylin	15 minutes.
3	Wash in tap water.	
4	Mordant in iron alum	1 minute.
5	Restain in alum haematoxylin	10 minutes.
6	Rinse in tap water.	
7	Stain in van Gieson's picro fuchsin	10 minutes.
8	Without washing differentiate in 95 per cent alcohol.	
9	Dehydrate in absolute alcohol.	
10	Clear and mount.	

Results

Haemoglobin: greenish-blue

Reference

DUNN, R.C., and THOMPSON, E.C. 1945, *Arch. Path.* **39**, 49.

HALE'S COLLOIDAL IRON METHOD

Solutions required

(a) Dialysed iron mixture
  Dialysed iron solution
  12.5 per cent aqueous acetic acid
 Mix equal parts immediately before use.
(b) Perl's reagent
  Potassium ferrocyanide in distilled water 2 per cent
  Hydrochloric acid in distilled water 2 per cent
 Mix equal parts immediately before use.
(c) Neutral red 1 per cent

Technique

1 Sections to water.
2 Place section in dialysed iron mixture   10 minutes.
3 Wash well in many changes of distilled water.
4 Flood slide with Perl's reagent   10 minutes.
5 Rinse in distilled water.
6 Wash in running water.
7 Counterstain lightly with neutral red $\frac{1}{2}$–1 minute.
8 Dehydrate, clear and mount.

Results

Acid mucopolysaccharides: blue
Nuclei: red
Other structures: shades of pink
Ferric iron if present: blue

Notes

(a) Sections must be thoroughly washed after treatment with the
 dialysed iron, to avoid the whole section becoming blue after
 treatment with Perl's reagent.
(b) It is simpler to buy the dialysed iron solution already prepared.

Reference

Hale, C.W. 1946, *Nature (Lond.)* **157**, 802.

HARRIS'S HAEMATOXYLIN

Solutions required

Haematoxylin	1 gm
Absolute alcohol	10 ml

Dissolve the haematoxylin in the alcohol with the aid of heat.

Aluminium potassium sulphate	20 gm
Distilled water	190 ml

Place the alum and distilled water in a large beaker; heat to dissolve. Mix the two solutions together and bring just to boiling point, stirring frequently.

Add mercuric oxide	0.5 gm

Cool by standing beaker in cold water.

When cool, add glacial acetic acid	0.8 ml

Technique

Use technique as detailed for Ehrlich's haematoxylin. Stain Harris's haematoxylin   5 minutes.

Notes

(a) Oxygen is liberated readily on addition of mercuric oxide, hence the need to use a wide-mouth beaker instead of a flask.

(b) Keeps well; store in a well-stoppered bottle.

Reference

HARRIS, H.F. 1900, *J. appl. Micr.* **3**, 777.

HEIDENHAIN'S IRON HAEMATOXYLIN

Solutions required

(a) Iron alum solution
   5 per cent ammonium ferric sulphate
(b) Haematoxylin solution

Haematoxylin	0.5 gm
Absolute alcohol	10 ml
Distilled water	90 ml

   It is best to keep the alcoholic haematoxylin as a stock solution
   and to add distilled water when required.

Technique

1 Section to water.
2 Treat section iron alum solution (see note (a)) $\frac{1}{2}$–24 hours.
3 Rinse in tap water.
4 Stain in haematoxylin solution (see note (b)) $\frac{1}{2}$–24 hours.
5 Rinse in tap water.
6 Differentiate in iron alum solution (see note (c))   1–30 minutes.
7 Wash in running tap water   10 minutes.
8 Counterstain if required.
9 Dehydrate, clear and mount.

Results

Muscle striations: blue-black
Nuclei: blue-black

Notes

(a) The iron alum solution is both the mordant and the differentiator.
(b) The section should be mordanted and stained for approximately the same length of time in each solution. Staining is complete when the section is overall black showing no detail.
(c) Differentiation is progressive, some structures being decolorised long before others; differentiation must therefore be carefully controlled using the microscope.

(d) It is important that all trace of iron alum be removed from the section if fading is to be prevented.

References

HEIDENHAIN, M. 1892, *Festschrift fur. Kolliker,* Leipzig.
HEIDENHAIN, M. 1894, *Archiv. f. mikr. Anat.* **43**, 423.
HEIDENHAIN, M. 1891, *Z. f. wiss. Mikr.* **9**, 198.
HEIDENHAIN, M. 1896, *Z. f. wiss. Mikr.* **13**, 186.

HUGHESDON'S METACHROMATIC METHOD FOR MUCIN

Solutions required

(a) 1 per cent potassium permanganate
(b) 5 per cent oxalic acid
(c) 0.2 per cent uranyl nitrate
(d) Staining solution, any one of the following:
   0.2 per cent azur A
   1 per cent azur B or C
   1 per cent toluidine blue, well ripened

Technique

1 Sections to water.
2 Oxidise section in 1 per cent potassium permanganate   1–5 minutes.
3 Rinse in tap water.
4 Bleach in 5 per cent oxalic acid until colourless.
5 Rinse in distilled water followed by rinse in tap water.
6 Stain in staining solution of choice   2 minutes.
7 Rinse in tap water.
8 Differentiate in 0.2 per cent uranyl nitrate   10 seconds or longer.
9 Rinse in tap water.
10 Blot section dry.
11 Complete dehydration in absolute alcohol.
12 Clear and mount.

Results

Mucin: metachromatically stained

Mast cell granules: red to violet

Other tissues: shades of blue

Note

Stage 2, oxidise sections to show epithelial mucin: 5 minutes; to show young connective tissue mucin: 1 minute.

Reference

HUGHESDON, P.E. 1949, *J. roy. micr. Soc.* **69**, 1.

MSB FOR FIBRIN

Solutions required

(a) Celestine blue

(b) Mayer's haemalum

(c) Martius yellow

Martius yellow	0.5 gm
Phosphotungstic acid	2 gm
95 per cent alcohol	98 ml

(d) Brilliant crystal scarlet 6R

Brilliant crystal scarlet 6R	1 gm
Acetic acid, glacial	2.5 ml
Distilled water	97.5 ml

(e) Soluble blue

Soluble blue	0.5 gm
Acetic acid, glacial	1 ml
Distilled water	99 ml

Technique

1 Sections to water.

2 Stain with celestine blue – haemalum sequence.

3 Rinse in 95 per cent alcohol.

4 Stain in martius yellow   2 minutes.

5 Rinse in distilled water.
6 Stain in brilliant crystal scarlet   10 minutes.
7 Rinse in distilled water.
8 Treat section with 1 per cent phosphotungstic acid to differentiate
  the red dye and to fix it in the fibrin   1–5 minutes.
9 Rinse in distilled water.
10 Stain in soluble blue   5–10 minutes.
11 Rinse in 1 per cent acetic acid.
12 Blot dry.
13 Complete dehydration in absolute alcohol, clear and mount.

Results
Nuclei: black
Red blood cells (erythrocytes): yellow
Fibrin: red
Fibrin 'old': bluish
Connective tissue: blue

Notes
(a) Primary or secondary fixation of the tissue or postchroming of
    the sections in a mercuric containing fixative will enhance the
    selectivity and brightness of the staining. The mercuric pigment
    should be removed before staining (see page 178).
(b) Thin paraffin sections are essential.
(c) This technique is selective for fibrin but not specific.
(d) Brilliant crystal scarlet 6R and soluble blue have other common
    names (ref. Conn.).

Reference
LENDRUM, A.C. et al. 1962, J. Clin. Path. **15**, 401.

MALLORY AND PARKER'S HAEMATOXYLIN

Solutions required

Haematoxylin	0.1 gm
2 per cent potassium dihydrogen phosphate	10 ml

Dissolve the haematoxylin in a few drops of absolute alcohol, mix with the 2 per cent potassium dihydrogen phosphate. Filter.

Technique

1 Section to water.
2 Stain in fresh haematoxylin mixture at 50–60 °C   2–3 hours.
3 Wash in running tap water   10 minutes.
4 Dehydrate, clear and mount.

Results

Lead: dark blue-grey
Copper: blue

References

MALLORY, F.B. 1938, *Pathological Technique,* Philadelphia: Saunders.
MALLORY, F.B., and PARKER, F. 1939, *Am. J. Path.* **15**, 517.

MALLORY'S PHOSPHOTUNGSTIC ACID HAEMATOXYLIN
(PTAH)

Solutions required

(a) PTAH

Haematin	1 gm
Phosphotungstic acid	20 gm
Distilled water	1000 ml

Dissolve the haematin and the phosphotungstic acid separately in distilled water, using gentle heat. When cool mix together and make up to 1000 ml.
Add 50 ml 0.25 per cent potassium permanganate to 'ripen'.

(b) 0.25 per cent potassium permanganate
(c) 5 per cent oxalic acid

Technique

1 Section to water.
2 0.25 per cent potassium permanganate   5 minutes.
3 Rinse in distilled water.
4 Bleach in 5 per cent oxalic acid   10 minutes.
5 Rinse in distilled water.
6 Wash in running water   5 minutes.
7 Rinse in distilled water.
8 Stain in PTAH   1–24 hours.
9 Dehydrate rapidly in absolute alcohol.
10 Clear and mount.

Results

Muscle striations
Fibrin
Nuclei
Fibroglia
Myoglia
Neuroglia fibres   light blue–dark blue
Centrioles
Erythrocytes
Intercellular bridges
Mitochondria
Myelin

Collagen, elastic fibres
Reticulin, cytoplasm   yellow to reddish brown
Cartilage
Bone matrix

Notes

(a) This technique is unique in that the single solution demonstrates a number of structures in two colours. The fact that a number of structures are demonstrated by one colour in no way detracts from its usefulness, as the colour is present in many shades and the structure identification is helped by its morphology.

(b) After PTAH do not go into water or the reds will be destroyed.
(c) Low-grade alcohols extract the reds.
(d) The length of staining time will determine the elements stained and degree of intensity.

References

MALLORY, F.B. 1897, *J. exp. Med.* **2**, 529.
MALLORY, F.B. 1900, *J. exp. Med.* **5**, 15.

MARSLAND, GLEES AND ERIKSON'S METHOD FOR AXONS

Solutions required

(a) 20 per cent silver nitrate in distilled water
(b) 10 per cent formalin in tap water
(c) 0.2 per cent gold chloride in distilled water
(d) 5 per cent sodium thiosulphate
(e) Ammoniacal silver solution
Mix together 30 ml of 20 per cent silver nitrate and 20 ml of 95 per cent alcohol. Add drop by drop 0.88 s.g. ammonia, shaking between each until solution is clear; then add 5 drops more. Filter before use.

Technique

1 Section to water.
2 Rinse in two changes of distilled water.
3 Transfer to 20 per cent silver nitrate at 37 °C
  (note (a))   30 minutes.
4 Rinse in distilled water.
5 Rinse in two changes of 10 per cent formalin   10 seconds each.
6 Wash off formalin with ammoniacal silver and
  flood slide   40 seconds.
7 Wash off ammoniacal silver with 10 per cent
  formalin and leave (note (b))   1 minute.
8 Rinse in distilled water.
9 Tone in 0.2 per cent gold chloride (note (c)) 3 minutes.
10 Wash in distilled water.
11 5 per cent sodium thiosulphate   5 minutes.
12 Wash in tap water   10 minutes.
13 Dehydrate, clear and mount.

Results

Axons and dendrites: black

Other tissues { (if toned): light grey
              { (untoned): yellowish-brown.

Notes

(a) Adjust time slightly, longer or shorter, depending when section is amber coloured.
(b) Examine section under the microscope; if the impregnation is incomplete, repeat stages 6 and 7. When satisfied with the result continue to stage 8.
(c) Sections need not be toned, depending on background colour required.

Reference

MARSLAND, T.H., GLEES, P., and ERIKSON, L.B. 1954, *J. Neuro-path. exp. Neurol.* **13**, 587.

MASSON-FONTANA FOR MELANIN AND ARGENTAFFIN

Solutions required
(a) 0.2 per cent gold chloride
(b) 5 per cent sodium thiosulphate
(c) 1 per cent neutral red
(d) Fontana's silver solution

To 5 per cent silver nitrate add ammonia (0.88 s.g.) drop by drop, agitating between each drop until the precipitate first formed just re-dissolves. Then add 5 per cent silver nitrate drop by drop, agitating between each drop until there is a persistent faint opalescence.

Technique

1 Wash section in distilled water, changing at 30-minute intervals	2 hours.
2 Place in a Coplin jar of Fontana's silver bath and keep in the dark	12–18 hours.
3 Wash in distilled water.	
4 Tone in 0.2 per cent gold chloride	5–10 minutes.
5 Wash in distilled water.	
6 5 per cent sodium thiosulphate	5 minutes.
7 Wash in water.	
8 Counterstain in 1 per cent neutral red	1 minute.
9 Rinse in water.	
10 Dehydrate, clear and mount.	

Results
Melanin, argentaffin pigment: black
Lipofuchsin and chromaffin pigment sometimes: greyish black
Nuclei: red
Other tissues: pinkish

Notes
(a) The silver solution may deposit black silver granules over the

section. It helps to put the section diagonally into the Coplin jar section downwards.

(b) Ammoniacal silver solutions if allowed to dry and deposit form an explosive compound. All glassware, therefore, should be carefully washed before and after use, and solutions should be freshly made and not stored.

References

MASSON, P. 1914, *C.R. Acad. Sci. (Paris)*, **158**, 57.
FONTANA, A. 1925–6, *Derm. Z.* **46**, 291.
FONTANA, A. 1912, *Derm. Wschr.* **55**, 1003.

MASSON'S TRICHROME STAIN
(Aniline blue or light green)

Solutions required

(a) Celestine blue

(b) Mayer's haemalum

(c) Masson's acid fuchsin

Acid fuchsin	1 gm
Glacial acetic acid	1 ml
Distilled water	100 ml

(d) Masson's aniline blue

Aniline blue	3 gm
Glacial acetic acid	2 ml
Distilled water	100 ml

Boil the distilled water and add the dye. Add the acetic acid, cool and filter.

(e) Masson's light green

Light green	1 gm
Glacial acetic acid	1 ml
Distilled water	100 ml

(f) 1 per cent phosphomolybdic acid

(g) 1 per cent glacial acetic acid aqueous

(h) 1 per cent glacial acetic acid alcoholic

143

Technique

1 Section to water.
2 Stain in celestine blue   5 minutes.
3 Rinse in water.
4 Stain in Mayer's haemalum   5–10 minutes.
5 Wash in running tap water for at least   10 minutes.
6 Rinse in distilled water.
7 Stain in acid fuchsin solution   5 minutes.
8 Rinse in distilled water.
9 Differentiate and mordant in a Coplin jar containing 1 per cent phosphomolybdic acid   5 minutes.
10 Without rinsing, place slide on cold staining rack and add just enough aniline blue solution to cover section   5 minutes.
11 Rinse in distilled water.
12 Differentiate in aqueous 1 per cent acetic acid to remove excess aniline blue. Control by frequent examination under the microscope.
13 Treat section in alcoholic 1 per cent acetic acid   30 seconds.
14 Dehydrate in absolute alcohol.
15 Clear and mount.

Results

Nuclei: blue-black
Cytoplasm: red
Collagen fibres: blue (or green)
Mucin: blue (or green)
Muscle: red

Notes

(a) Light green may be substituted in stage 10 if colour preferred.
(b) Aniline blue or light green is applied to section without rinsing as the (1 per cent) phosphomolybdic acid acts as a mordant.

Reference

MASSON, P. 1928, *Amer. J. of Path.* **4**, 181.

MAYER'S HAEMALUM

Solutions required

Haematoxylin	1 gm
Distilled water	1000 ml

Dissolve with the aid of gentle heat.

Aluminium ammonium sulphate	50 gm
Sodium iodate	0.2 gm

Shake until the alum is dissolved.

Add citric acid	1 gm
Chloral hydrate	50 gm

The colour turns reddish-purple. Leave seven days before using.

Technique

1 Section to water.
2 Stain in Mayer's haemalum   5–10 minutes.
3 Wash in running tap water.
4 Examine under the microscope: if staining is correct, continue to wash in running tap water to ensure complete removal of acid content of staining solution.
5 Counterstain as required.
6 Dehydrate, clear and mount.

Results

Nuclei: blue
Other tissues as counterstained

Notes

(a) 5–10 minutes; this depends on the age of the solution – determine by test staining. Use progressively (i.e. not overstaining).
(b) As this is a progressive nuclear stain it is possible to use it after another stain has been applied to demonstrate the main tissue element. (The nuclear stain becomes the counterstain.)

Reference

MAYER, P. 1891, *Mitt. Zool. Stat. Neapel.* **10**, 170–186.

METACHROMASY

The choice of a particular metachromatic dye depends on the tissue element to be demonstrated. This staining method has the advantage of being easy to do and the results quick to see. The disadvantages are the leaching out of the dye into the watery mountant, as almost without exception the preparation of a 'permanent' mount is difficult.

Metachromasy or metachrosis: A change of colour.

Metachromasia or metachromatism: The property of tissue to become stained in different tints by the same stain.

Metachromatic: A single dye is metachromatic if it demonstrates specific tissue elements a different tint from the other tissues.

Technique
1 Section to water.
2 Stain in selected solution.
3 Rinse in distilled water.
4 Differentiate if required.
5 Rinse in water.
6 Mount in watery mountant.

Results
Metachromatic substances: red, pink or purple
Nuclei and other tissues: blue

Notes
(a) Mucin–toluidine blue 1 per cent   10–60 seconds.
(b) Amyloid–methyl violet 1 per cent   3–5 minutes.
(c) Differentiation if required for (a) is performed in tap water, or for (b) in 1 per cent glacial acetic acid.

MYELIN SHEATHS:
WEIL'S METHOD FOR PARAFFIN SECTIONS

Solutions required
(a) 4 per cent ferric ammonium sulphate in distilled water (iron alum)
(b) 1 per cent haematoxylin in 10 per cent alcohol
(c) Iron haematoxylin solution
     Mix equal parts (a) and (b) together as required.
(d) Weigert's differentiator

Borax	1 gm
Potassium ferricyanide	1.25 gm
Distilled water	100 ml

Technique
1 Sections to water.
2 Stain in Coplin jar of iron haematoxylin at 37 °C   30 minutes.
3 Rinse in water.
4 Differentiate in 4 per cent iron alum until end point is reached. (Control differentiation with frequent use of the microscope.) Background yellowish-brown.
5 Rinse in distilled water.
6 Complete differentiation in Weigert's differentiator.
7 Wash well in running tap water.
8 Dehydrate, clear and mount.

Results
Myelin sheaths: blue-black
Nuclei: greyish
Background: colourless
Erythrocytes: black

Reference
WEIL, A. 1928, *Arch. Neurol. Psychiat. (Lond.)* **20**, 392.

NILE BLUE SULPHATE

Solutions required
(a) 2 per cent Nile blue sulphate
(b) 0.1 per cent acetic acid
(c) Glycerine/gelatine

Technique
1 Frozen section to distilled water.
2 Stain in a covered dish of Nile blue sulphate   20 minutes.
3 Rinse in distilled water.
4 Differentiate in 0.1 per cent acetic acid until neutral lipids pink and acidic lipids blue.
5 Wash well in water.
6 Mount in glycerine jelly.

Results
Neutral lipids: pink-red
Acidic lipids: blue

Notes
(a) Check that oxazone is present in the dye sample by adding a few grains of the sample to a little xylene; if oxazone is present the red component will show immediately.
(b) This method cannot be regarded as specific. It is generally agreed that the neutral lipids are stained pink, but the specificity of the blue staining is in question.

Reference
CAIN, A.J. 1947, *Anat. J. micr. Sci.* **88**, 383.

NISSL BODIES

Solution required

Thionin solution

1 per cent thionin in distilled water	1 ml
pH 3.6 acetate buffer	39 ml

Technique

1 Section to distilled water.
2 Stain in thionin solution   30 minutes.
3 Brief differentiation in 70 per cent alcohol.
4 Dehydrate, clear and mount.

Results

Nissl bodies: bright blue
Nuclei: light blue
Cytoplasm and background: colourless

Note

Differentiations must be carefully controlled.

Reference

NISSL, F. 1885, *Neurol. Z. bl.* **4**, 500.

ORCEIN FOR ELASTIC FIBRES

Solution required

Orcein solution

Orcein	1 gm
70 per cent alcohol	100 ml
Hydrochloric acid	1 ml

Dissolve the orcein in the alcohol, using gentle heat, cool and filter, add the hydrochloric acid. Keeps indefinitely.

Technique

1 Sections to alcohol.
2 Stain in orcein at 37 °C   30 minutes to 2 hours.
   (Control staining time by examining section with a microscope at intervals.)
3 Wash well in water.
4 If counterstaining, remove background staining with 1 per cent acid alcohol.
5 Wash well in distilled water.
6 Counterstain if required.
7 Dehydrate, clear and mount.

Results

Elastic fibres: dark brown
Other tissues: as counterstained

Note

Counterstains that may be used are haematoxylin, neutral red and van Gieson's picro fuchsin. Care must be taken not to obscure the finer fibres by using a non-contrasting counterstain or by counterstaining too heavily.

Reference

UNNA, P. 1894, *Mischr. prakt. Dermatol.* **19**, 397.

OXYTALAN FIBRES

Solutions required
(a) 15 per cent potassium peroxymonosulphate
(b) Orcein solution

Technique

1 Sections to water	30 minutes.
2 Oxidise in potassium peroxymonosulphate	5 minutes.
3 Wash in running tap water.	
4 Transfer to 70 per cent alcohol.	
5 Immerse in a covered Coplin jar of orcein solution at 37 °C	30 minutes to 2 hours.
6 Wash in running tap water.	
7 Dehydrate, clear and mount.	

Results
Oxytalan fibres: dark brown

References
FULLMER, H.M. *et al.* 1958, *J. Histochem. and Cytol.* **6**(6), 425–30.
RANNIE, I. 1963, *Trans. Euro. Orthodon. Soc.,* 1–10.

PERIODIC ACID SCHIFF TECHNIQUE

Solutions required
(a) 0.5 per cent periodic acid
(b) Mayer's haemalum
(c) Sulphurous acid

Sodium metabisulphite 10 per cent	6 ml
N/1 hydrochloric acid 10 per cent	5 ml
Distilled water	100 ml

(d) Schiff's reagent

Basic fuchsin	1 gm
Sodium metabisulphite, anhydrous	1 gm
Distilled water	200 ml
N/1 hydrochloric acid	20 ml

Boil the distilled water, add basic fuchsin and stir, cool to 50 °C. Then filter and add hydrochloric acid, cool to 25 °C and add the sodium metabisulphite.

This solution is ready for use when it becomes nearly colourless, which may take up to two days in the dark.

(Alternatively activated charcoal may be added to the solution, shaken and filtered. The solution is then ready for use.) When the solution becomes recoloured it should be discarded.

Technique
1 Section to water.

2 Periodic acid 0.5 per cent	5 minutes.
3 Rinse in distilled water.	
4 Schiff's reagent	15 minutes.
5 Rinse in three fresh changes of sulphurous acid, 2 minutes in each change	6 minutes.
6 Wash in running tap water	5 minutes.
7 Counterstain in Mayer's haemalum	30 seconds.
8 Wash in running tap water	5 minutes.
9 Dehydrate, clear and mount.	

Results
Positive material: reddish-purple
Nuclei: faint grey

Notes
(a) Among the many substances both normal and abnormal that give a positive reaction are the following commonly seen ones: Glycogen, mucin, reticulin, basement membrane, thyroid colloid, amyloid, arteriolosclerotic hyaline, fungi and others.
(b) As many substances are demonstrated by this technique the identification of the substance should be confirmed by other techniques if possible.
(c) The counterstaining should only ghost in the nuclei, so as not to mask the appearance of positive reacting material.
(d) The authors believe that when preparing Schiff's reagent decolorisation of the solution is best achieved by placing in the dark for two days. The alternative method is faster, but could cause negative staining.

Reference
McMANUS, J.F.A. 1946, *Nature (Lond.)* **158**, 202.

PERLS' PRUSSIAN BLUE REACTION
(Free iron)

Solutions required
(a) 2 per cent hydrochloric acid (A.R. grade)
(b) 2 per cent potassium ferrocyanide in iron-free distilled water.
(c) 1 per cent eosin Y

Technique
1 Section to water and place on cold staining rack.
2 Mix equal parts 2 per cent hydrochloric acid and 2 per cent potassium ferrocyanide in a test tube and warm (note (b)), pour over the section   10 minutes.
3 Wash in distilled water.
4 Counterstain in 1 per cent eosin   2 minutes.
5 Differentiate in tap water until lightly stained.
6 Dehydrate, clear and mount.

Results
Haemosiderin/ferric salt/free iron: dark blue
Connective tissue: shades of pink
Other pigments: unstained

Notes
(a) Avoid contact of the tissue or section with iron or acid containing fluid except as indicated in the technique.
(b) Do not overheat the mixture (above 40 °C).
(c) A nuclear contrasting counterstain may be used if desired.

Reference
PERLS, M. 1867, *Virchows Arch. path. Anat. Physio.* **39**, 42.

PHLOXINE TARTRAZINE

Solutions required
(a) Mayer's haemalum
(b) Phloxine

Phloxine	0.5 gm
Calcium chloride anhydrous	0.5 gm
Distilled water	100 ml

Mix together. This solution keeps well.
(c) Tartrazine
Prepare a saturated solution of tartrazine in cellosolve (2–Ethoxyethanol) approx. 0.3 per cent

Technique
1 Sections to water.
2 Stain in Mayer's haemalum   8 minutes.
3 'Blue' in running tap water.
4 Stain in phloxine mixture in a Coplin jar   30 minutes.
5 Rinse in water and blot almost dry.
6 Counterstain and differentiate in the tartrazine solution (controlling microscopically)   5–15 minutes.
7 Rinse and dehydrate in absolute alcohol.
8 Clear and mount.

Results

Erythrocytes
Inclusion bodies
Paneth cell granules
Russell bodies   } red
Fibrin
Keratin
Pancreatic beta cells

Nuclei: grey-blue
Other tissues: yellow

Reference
LENDRUM, A.C. 1947, *J. Path. Bact.* **59**, 399.

RHODAMINE B FOR KERATIN

Solutions required
(a) Toluidine blue solution
      Toluidine blue   0.1 gm
      Distilled water   100 ml
(b) Acetic acid–sodium acetate
     buffer pH 3.6 (Warpole)
      $N/10$ acetic acid   185 ml
      $M/10$ sodium acetate pH 3.6   15 ml
(c) Rhodamine B solution
      Rhodamine B   0.1 gm
      Buffer solution   100 ml

Technique
1 Sections to water.
2 0.1 per cent toluidine blue   10 minutes.
3 Rinse in 3 changes distilled water.
4 0.1 per cent rhodamine solution   10 minutes.
5 Rinse in 3 changes of distilled water.
6 Dehydrate, clear and mount.

Results
Keratin: bright red
Nuclei and all basophilic material: blue or blue-black

Note
May be viewed under ultraviolet light if desired.

Reference
LIISBERG, M.F. 1968, *Acta. Anat.* **69**, 52–7.

SCHMORL'S PICRO THIONIN

Solutions required

(a) Nicolle's carbol thionin

Thionin	1 gm
Phenol	1 gm
80 per cent ethyl alcohol	10 ml
Distilled water	100 ml

Using a pestle and mortar, grind the thionin into the alcohol, dissolve the phenol in the distilled water and add to the thionin solution.

Keeps as a stock solution. Just before use add 1–2 drops of 0.88 s.g. ammonia to each 5 ml of stock solution and filter.

(b) Picric acid

Prepare a saturated solution of picric acid by dissolving picric acid to excess in hot distilled water, allow to cool and filter.

Technique

1 Sections to distilled water.

2 2 changes of distilled water during   10 minutes.

3 Transfer sections to ammoniacal carbol thionin
(see note (b)) $\frac{1}{2}$–10 minutes.

4 Rinse in distilled water   20 seconds.

5 Transfer section to fresh picric acid $\frac{1}{2}$–1 minute.

6 Differentiate in 70 per cent alcohol till stain
ceases to flood from the section   1–10 minutes.

7 Rinse in distilled water.

8 Treat in picric acid solution (see note (c))   1 minute.

9 Rinse in distilled water.

10 Dehydrate in 96 per cent alcohol.

11 Clear in origanum oil or carbol xylol.

12 Mount in colophonium turpentine.

Results

Lacunae; canaliculi; dentinal tubules and lateral branches: dark brown–black

Background: yellow–yellow-brown

Cartilage; nuclei: purplish-red

Notes

(a) Celloidin or frozen sections 15 microns or more thick should be used, but reasonable results may be obtained using paraffin sections cut and mounted as thick as possible.

(b) If the carbol thionin solution is made too alkaline by the addition of too much ammonia, crystals will form on the section.

(c) Treatment of the section with picric acid in stage 8 restores the required colour to the bone or tooth matrix sometimes lost in the 70 per cent alcohol and water treatment.

(d) Prolonged dehydration will result in over-differentiation of the matrix.

(e) When using carbol xylene, protective gloves should be worn in a well-ventilated environment.

(f) This is not a true staining method but a precipitation of the dye into spaces within calcified tissues. It may be applied to ground sections of undecalcified tissue or to sections of decalcified tissue. Celloidin or frozen sections are preferred to paraffin sections as there is little or no shrinkage of the spaces and thicker sections can be cut. However, paraffin sections will probably be the most readily available, and with care this technique will give useful results.

References

SCHMORL, G. 1899, *Centr. f. allg. Pa. h.* **10**, 745.

SCHMORL, G. 1900, *Vehandl. d. Naturforsher Ges. Muncher II,* **2**, 71.

SCHMORL, G. 1921, *Die Pathologisch-Histoligischem Untersuchungsmetoden.*

SILVER IMPREGNATION FOR
BASEMENT MEMBRANE

Solutions required
(a) 0.5 per cent periodic acid
(b) 0.2 per cent gold chloride
(c) 3 per cent sodium thiosulphate
(d) Methenamine silver solution

3 per cent methenamine	100 ml
5 per cent silver nitrate	5 ml

Mix and shake immediately, continue shaking until white precipitate dissolves. Keep at 4 °C as stock solution.

Take stock methenamine silver solution	50 ml
5 per cent sodium tetraborate (borax)	6 ml

Mix fresh as required.

Technique
1 Sections to water.
2 Rinse in distilled water.
3 Periodic acid   15 minutes.
4 Wash in distilled water.
5 Silver solution heated and maintained at 50 °C
   in the dark $1\frac{1}{2}$ to 3 hours.
6 Wash in distilled water.
7 Tone in gold chloride   2 minutes.
8 Wash in distilled water.
9 Sodium thiosulphate   2 minutes.
10 Running tap water   5 minutes.
11 Counterstain with haematoxylin and eosin if required.
12 Dehydrate, clear and mount.

Results
Basement membrane: black

Notes

(a) The best results are obtained using sections less than 4 micrometers thick.

(b) Care must be taken to ensure that the silver solution is warmed to 50 °C before placing the section in it and that the temperature is evenly maintained at this during the whole of the time.

(c) After 1½ hours the section should be rinsed in distilled water and examined microscopically and repeated at 30-minute intervals up to 3 hours until the impregnation is satisfactory.

Reference

JONES, D.B. 1957, *Am. J. Path.* **33**, 313.

SILVER IMPREGNATION FOR FUNGI
(Grocott's Mod. Gomori)

Solutions required

(a) 5 per cent aqueous sodium tetraborate (borax)

(b) 5 per cent silver nitrate in distilled water   5 ml

(c) 3 per cent methenamine   100 ml
    Mix silver nitrate and methenamine together to make stock silver solution.

(d) 5 per cent chromium trioxide

(e) 1 per cent sodium bisulphite

(f) 0.1 per cent gold chloride

(g) 2 per cent sodium thiosulphate

(h) 2 per cent light green in 1 per cent acetic acid

(i) Working silver solution
        5 per cent borax   2 ml
        Distilled water   25 ml
    Mix and add 25 ml stock silver solution.

Technique

1 Sections to water.
2 5 per cent chromium trioxide   1 hour.
3 Wash in running water   10 seconds.
4 Rinse in 1 per cent sodium bisulphite.
5 Wash in tap water   5 minutes.
6 Wash in distilled water, 4 changes.
7 Methenamine silver solution at 50 °C in the dark
   (see note (a))   30–60 minutes.
8 Rinse in 4 changes distilled water.
9 Tone in 0.1 per cent gold chloride   2.5 minutes.
10 Rinse in distilled water.
11 2 per cent sodium thiosulphate   2.5 minutes.
12 Wash well in water.
13 Counterstain 2 per cent light green solution   30–60 seconds.
14 Dehydrate, clear and mount.

Results

Fungi 'walls' outlined black.

Notes

(a) Ensure that the solution is at working temperature before using.
(b) Reticulin fibrils and fibrin may be blackened and must not be mistaken for fungi.

References

GOMORI, G. 1946, *Amer. J. clin. Path.* **16**, 177.
GROCOTT, R.G. 1955, *Amer. J. clin. Path.* **25**, 975.

SOUTHGATE'S MODIFICATION OF MAYER'S MUCICARMINE METHOD

Solutions required
(a) Ehrlich's haematoxylin
(b) Southgate's mucicarmine

Carmine	1 gm
Aluminium hydroxide	1 gm
Anhydrous aluminium chloride	0.5 gm
50 per cent alcohol	100 ml

Place carmine in a 500-ml flask, add the aluminium hydroxide and alcohol, shake well to dissolve and mix. Add the aluminium chloride. Stand flask in a waterbath and bring to boil quickly; boil for $2\frac{1}{2}$ minutes. Stand flask in cold running water to cool solution quickly; filter.

Technique
1 Sections to water.
2 Stain in Ehrlich's haematoxylin   5 minutes.
3 Differentiate and 'blue' as for H and E.
4 Stain in mucicarmine diluted with distilled water
  to one-third strength   30 minutes.
5 Rinse in water.
6 Dehydrate, clear and mount.

Results
Mucin: red
Nuclei: blue

Notes
(a) Mucicarmine solution keeps well as a stock solution but is unstable when diluted; the one in three solution should therefore be prepared only as required.
(b) Epithelial mucin stains well; connective poorly if at all.

References
MAYER, P. 1896, *Mitt. Zool. Stat. Neapel,* **12**, 303.
SOUTHGATE, H.W. 1927, *J. Path. and Bact.* **30**, 729.

162

MODIFIED STEEDMAN'S ALCIAN BLUE METHOD

Solutions required

(a) Alcian blue solution

1 per cent aqueous alcian blue 8GS	50 ml
1 per cent acetic acid	50 ml
Filter then add thymol	10 mg

(b) 1 per cent aqueous neutral red

Technique

1 Sections to water.
2 Stain in alcian blue   10 minutes.
3 Rinse in water.
4 Counterstain in neutral red   2 minutes.
5 Commence differentiation in water.
6 Complete differentiation in alcohol, while dehydrating.
7 Clear and mount.

Results

Mucin, mast cell granules
Ground substance cartilage: blue–blue-green
Nuclei: red

Notes

(a) If colour is preferred alcian green may be substituted.
(b) This simple test may be used in a more complex way and reference should be made to the textbooks.

References

STEEDMAN, H.F. 1950, *Qt. J. Microsc. Sci.* **91**, 477.
LISSON, L. 1954, *Stain Technol.* **29**, 131.

SUDAN III AND IV IN HERXHEIMER'S SOLUTION

Solutions required
(a) Mayer's haemalum
(b) Sudan mixture
   To 100 ml of acetone add equal parts Sudan III and IV dry dye powder to produce supersaturated solution, allow to stand in a tightly stoppered bottle for a few hours, then add 100 ml of 70 per cent alcohol. Store in tightly stoppered bottle.
   Filter before use.

Technique
1 Frozen sections to 70 per cent alcohol.
2 Stain in Sudan mixture (note (a))   10 minutes.
3 Transfer section to 70 per cent alcohol to differentiate excess stain from section.
4 Wash in water.
5 Stain nuclei in Mayer's haemalum lightly   2 minutes.
6 Mount on slide, drain water from slide but do not allow section to dry, mount in glycerine gelatine.

Results
Neutral fats: red
Myelin lipids: pink
Fatty acids: orange-yellow
Nuclei: light grey-blue

Notes
(a) The Sudan solution must be used in a closed vessel to avoid evaporation and precipitation.
(b) The sections are stained floating free.
(c) The coverslip on the finished preparation should have its edges sealed with a sealant such as clear nail varnish.

Reference
HERXHEIMER, G.W. 1903, *Zbl. allg. Path. path. Anat.* **312**, 405.

SUDAN BLACK

Solution required
Sudan black solution
    Sudan black B   7 gm
    70 per cent alcohol   500 ml
Add the dye to the alcohol and warm to 56 C, maintain at this temperature for 1 hour. When cool, store in a tightly stoppered bottle.
Filter before use.

Technique
1 Sections to 70 per cent alcohol.
2 Stain in Sudan black (note (a))   10 minutes.
3 Rinse in 70 per cent alcohol.
4 Wash in water.
5 Mount in glycerine gelatine.

Results
Lipids: black
Other tissue: greyish

Note
Staining solution must be used in tightly closed vessel.

Reference
LISON, L., and DAGNELIE, J. 1935, *Bull. Histol. app. Physical Path.* **12**, 85.

UNNA-PAPPENHEIM, TREVAN AND SHARROCK MODIFICATION

Solutions required

(a) Stock solutions

Methyl green

Prepare a 2 per cent aqueous solution of methyl green and wash solution repeatedly with chloroform until chloroform remains clear.

(b) 5 per cent pyronin Y

(c) M/5 acetate buffer pH 4.8

0.1 M sodium acetate	119 ml
0.1 M acetic acid	81 ml

Working solutions

2 per cent methyl green	10 ml
5 per cent pyronin	17.5 ml
Distilled water	250 ml

Plus equal part of solution (c)

Technique

1 Sections to distilled water and blot dry.

2 Stain in working solution   20–30 minutes

3 Rinse rapidly in distilled water, blot until almost dry.

4 Dehydrate rapidly in acetone.

5 Clear in xylene and mount.

Results

DNA: green–bluish green

RNA: red–reddish purple

Note

There are numerous reasons why this technique may not work. Reference must be made to the more detailed technique textbook.

Reference

TREVAN, D.J., and SHARROCK, A. 1951. *J. Path. Bact.* **63**, 326.

van GIESON'S PICRO FUCHSIN TECHNIQUE

Solutions required

(a) Picro fuchsin solution

Saturated sol. of picric acid in
distilled water   90 ml
1 per cent acid fuchsin in
distilled water   10 ml

Mix together and boil for three minutes and allow to cool before use. This solution keeps and a stock bottle ready for use should be maintained.

(b) Celestine blue

Celestine blue B   0.5 gm
Ferric ammonium sulphate   5 gm
Glycerine   14 ml
Distilled water   100 ml

Without heat dissolve the ferric ammonium sulphate in the distilled water. Add the celestine blue and boil for three minutes in a wide-mouth glass vessel. Cool and filter, then add the glycerine. This solution keeps for approximately six months.

(c) Mayer's haemalum (**page 145**).

Technique

1 Sections to water.
2 Stain in the celestine blue solution   5 minutes.
3 Rinse in water.
4 Stain in Mayer's haemalum   5 minutes.
5 Wash in running tap water for at least   5 minutes.
6 Stain in the picro fuchsin solution   5 minutes.
7 Rinse rapidly in distilled water.
8 Dehydrate rapidly in absolute alcohol.
9 Clear and mount.

Results

Collagen: ranging from pink to red
Muscle and red blood cells (erythrocytes): yellow
Nuclei: blackish

Notes

(a) Care must be taken with stages 7 and 8 as extraction and differentiation of the acid fuchsin occurs in water and of the picric acid in alcohol.

(b) Fading of the collagen stain will take place in time, but restaining is possible.

(c) Picric acid differentiates haematoxylin, therefore the more resistant celestine blue–iron alum sequence or an iron haematoxylin must be used.

References

VAN GIESON, I. 1889, *N.Y. St. Med.* **50**, 57.

LENDRUM, A.C., and MCFARLANE, 1940, *J. Path. Bact.* **50**, 381.

VERHOEFF'S HAEMATOXYLIN FOR ELASTIC FIBRES

Solutions required

(a) 5 per cent haematoxylin, alcoholic solution

(b) 10 per cent ferric chloride

(c) Lugol's iodine
These may be kept as stock solutions, although the haematoxylin solution should not be used if more than a few weeks old.

(d) Verhoeff's haematoxylin (mix fresh for use)

Haematoxylin	20 ml
Ferric chloride	8 ml

Mix together and shake quickly.

Lugol's iodine	8 ml

Add to mixture and shake quickly.

Technique

1 Section to absolute alcohol.
2 Stain in Verhoeff's haematoxylin   15–30 minutes.
3 Rinse in water.
4 Differentiate in ferric chloride 2 per cent (control differentiation under the microscope).
5 Immerse in alcohol 95 per cent to remove the iodine   5 minutes.
6 Wash in water.
7 Counterstain with eosin 1 per cent or van Gieson's picro fuchsin.
8 Dehydrate, clear and mount.

Results

Elastic fibres: black
Nuclei: blackish
Other tissue as counterstained.

Notes

(a) Differentiation must be carefully controlled so that the finer fibres are not lost. This technique is best reserved for the demonstration of coarse fibres.
(b) If counterstaining with eosin the differentiation of the haematoxylin must be to the required picture; if using picro-fuchsin the haematoxylin should be slightly under-differentiated, as the acid in the counterstain will complete the differentiation.

Reference

VERHOEFF, E.H. 1908, *J. Amer. med. Ass.* **50**, 876.

von KOSSA'S SILVER METHOD FOR CALCIUM

Solutions required
(a) 5 per cent silver nitrate
(b) 5 per cent sodium thiosulphate
(c) 1 per cent neutral red
(d) Saturated aqueous lithium carbonate

Technique
1 Sections to water.
2 Treat sections with saturated solution of lithium carbonate to remove urates if present   20 minutes.
3 Wash well in running tap water.
4 Wash well in distilled water.
5 Place section in 5 per cent silver nitrate under direct bright light
$\frac{1}{2}$–1 hour.
6 Wash in distilled water.
7 Treat section with 5 per cent sodium thiosulphate   5 minutes.
8 Wash in running tap water.
9 Counterstain in 1 per cent neutral red   1 minute.
10 Wash in water.
11 Dehydrate, clear and mount.

Results
Calcium salts: dark brown-black
Nuclei: red

Notes
(a) Acid containing fixative should be avoided.
(b) Any bright light will do.

Reference
VON KOSSA, J. 1901, *Ziegler's Beitr.* **29**, 163.

WADE–FITE METHOD FOR AFB

Solutions required
(a) Carbol fuchsin

Basic fuchsin	1 gm
Absolute alcohol	10 ml
5 per cent aqueous phenol	100 ml

Dissolve the basic fuchsin in the alcohol, then add the 5 per cent phenol.
(b) Mayer's haemalum
(c) 10 per cent sulphuric acid

Technique
1 Remove wax from section with xylene.
2 Blot dry.
3 Wash in water   5 minutes.
4 Stain in carbol fuchsin at 21 °C   30 minutes.
5 Decolorise in 10 per cent sulphuric acid.
6 Wash in water.
7 Counterstain with Mayer's haemalum   10 minutes.
8 Wash in water   10 minutes.
9 Blot dry and complete drying in 56 C incubator.
10 Clear in xylene and mount.

Results
AFB: red
Nuclei: blue

Reference
Azulay, R.D., and Andrade, L.M.C. 1954, *Int. J. Leprosy*, **22**, No. 2, 195.

ZIEHL NEELSEN

Solutions required

(a) 1 per cent hydrochloric acid in 70 per cent alcohol
(b) 1 per cent methylene blue
(c) Carbol fuchsin

Basic fuchsin	1 gm
Absolute alcohol	10 ml
5 per cent phenol in water	100 ml

Dissolve the basic fuchsin in the alcohol, then mix with the phenol solution.

Technique

1 Sections to water.
2 Place section on cold staining rack and filter the carbol fuchsin on to the section, flooding the whole of the slide. Heat from below until the solution begins to steam   5–10 minutes.
3 Rinse in water.
4 Decolorise the section using acid alcohol until section is almost colourless   1–3 minutes.
5 Rinse in water.
6 Counterstain in stock methylene blue diluted to 0.1 per cent   1 minute.
7 Rinse in water.
8 Differentiate the counterstain until pale blue and dehydrate in absolute alcohol.
9 Clear and mount.

Results

Acid-alcohol fast bacilli: red
Other bacteria: blue
Other tissues: blue
Erythrocytes: reddish

Notes

(a) Take care not to overheat the solution or to let the slide dry up. Use a pledget of cotton wool dipped in alcohol, hold in forceps and ignite to warm the section.

(b) A control section known to contain AAFB should always be stained simultaneously.

References

ZIEHL, F. 1882, *Dt. med. Wschr.* **8**, 451.
NEELSEN, F. 1883, *Zentbl. med. Wiss.* **21**, 497.

PART 3

Common processing and sectioning methods

TABLE OF PROCESSING POSSIBILITIES

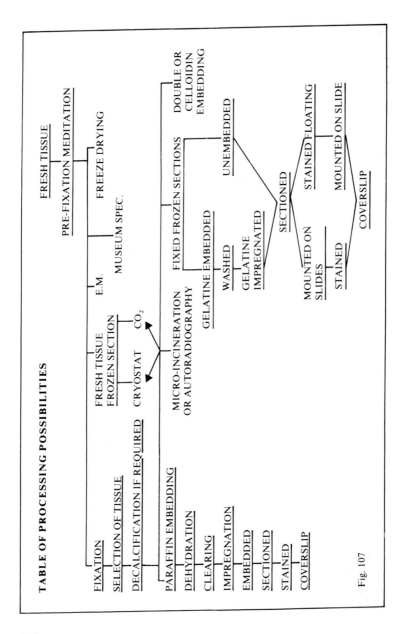

Fig. 107

TISSUE-PROCESSING PROCEDURES

PRE-FIXATION

Ideally, while the specimen is still fresh, consider for what purpose the specimen is required. Is there any chance that it may be required for museum display and is the preservation of macro-anatomical and micro-anatomical relationships therefore of equal importance? Or can one take precedence over the other? For example, if a lung is re-inflated by being perfused with excessive pressure, the delicate tissue structures will rupture, destroying the microscopical picture. Or the packing of intestine with cotton wool to preserve the tube effect will, if the pressure is excessive, flatten the mucosa, thereby losing the macro-anatomical picture and disturbing the microscopic appearance. The specimen should therefore be fixed in as lifelike a position as possible. The introduction of mechanical artefacts must be avoided, for example by not clamping the area of interest or producing a 'sinus' with a needle.

The counsel of perfection is that the specimen, or selected pieces, should invariably be placed in the fixative recommended for the tissue element to be demonstrated. In practice, this is rarely possible except under controlled research conditions.

PROCESSING

RULES

(a) The processing method used must depend on the tissue element to be demonstrated. For example, it is no use processing the tissue to paraffin wax if you wish to demonstrate fat, but you can process on to gelatine.

(b) Each stage of processing must be adequate in itself to produce a well-processed piece of tissue.

(c) All stages of fixing and processing must be carried out in adequate quantities of the reagents; as a rough guide, one hundred times the volume of the tissue.

FIXATION

Fix: to make firm, secure or fastened.

Fixation: in process of fixing, rendering solid, coagulate or arrest development.

Fixative: substance used in the process of fixation.

Fixed: 'permanently' preserved in relative position.

Tissue is usually fixed before processing and sectioning to preserve the tissue constituents in as life-like a manner as possible with maintained relationship and 'firm' it to a consistency suitable for sectioning. Fixation also helps the tissue to resist distortion by shrinkage or swelling during subsequent processing.

Methods of fixation

1 Immersion of tissue in the fixative.
2 Suspension of tissue in the fixative vapour (not in contact with fixative).
3 Perfusion of the specimens before or after removal from the body to facilitate penetration of the fixative to ensure rapid and even fixation.

Primary fixation is the first fixative used, not always of choice but dictated by local requirements.

Secondary fixation is treatment of the selected piece of primary fixed tissue in a fixative of choice in an attempt to obtain the benefits of this fixative.

Post-chroming is treatment of the section in a second fixative to try and influence the end staining results.

4 After primary or secondary fixation in any solution containing mercuric chloride, it is necessary to remove the fixation 'pigment' before staining the section.
   (a) Sections to absolute alcohol.
   (b) Treat with 0.5 per cent iodine in 70 per cent alcohol – 5 minutes.
   (c) Rinse in tap water.
   (d) Treat with 2.5 per cent sodium thiosulphate until bleached – 1 to 2 minutes.
   (e) Wash in running tap water.
   (f) Follow staining technique instructions.

Factors influencing choice of fixative

1 Thickness of gross tissue compatible with the penetrating power of fixative.
2 Possibility of multiple slicing of the fresh tissue to allow the use of more than one fixative.
3 Possible use as a museum specimen.
4 Tissue constituent(s) to be demonstrated.
5 Mordanting effect, related to staining techniques required.
6 Type of tissue, vascular allowing better penetration than avascular.
7 Time factor (diagnostic material usually required faster than research material).

Type of fixative

1 Simple: single reagent.
2 Compound: mixtures, whose components are claimed to cancel out each others' harmful effects, leaving only the most beneficial.
3 Micro-anatomical: preserved relationship of the tissue of primary importance.
4 Cytological: (A) nuclear, preservation of nuclei components.
   (B) cytoplasm, preservation of cytoplasm components.
5 Histochemical.

In practice the choice of a fixative and the method of fixation is frequently a compromise. Usually, the method of fixation is by immersion of the gross specimen, however large, this procedure being carried out away from the laboratory and by non-technical staff. For a variety of reasons, some time may elapse before the specimen is received by the histologist. The most frequent requirement of fixation is that the 'gross' microscopical tissue constituents and relationships are preserved. A 'safe' general-purpose macro-anatomical and micro-anatomical fixative is 10 per cent formol saline. Immersion fixation of the gross specimen is usual; it is relatively safe to handle, damages the tissue less than most other fixatives, the fixation time is not critical, it permits secondary fixation, the use of a wide variety of processing and staining techniques and in some instances, as with lipids, does not fix the tissue constituent, but preserves it for subsequent treatment.

, After fixation, pieces of tissue are selected from the specimen for sectioning by the required method. If secondary fixation is indicated the previously fixed piece of tissue or unattached frozen section may be treated with the fixative of choice and to a large extent is usually successful. Tissue sections adhering to a glass slide cannot be secondarily fixed; they are postchromed, only limited benefit being obtained.

DECALCIFICATION

Decalcify: to remove calcium from calcified tissues.

If calcified tissue is to be sectioned it is almost invariably decalcified after fixation and before processing to an embedding medium. The ideal decalcifying method would satisfy the following criteria:

(a) Inexpensive.
(b) Simple to use.
(c) Avoid all damage to tissue.
(d) Would not inhibit the use of staining or impregnation technique.
(e) Little or no delay in processing of tissue.

At best it is possible to achieve only some of these requirements with any one of the available methods, each having its limitations.

Methods of decalcification

(a) Simple, single dilute acid solution.
(b) Compound mixture of chemicals and dilute acids.
(c) Ion-exchange resin.
(d) Electrolytic.
(e) Chelation.

The most frequently used method is the simple, easy to use single dilute acid solution; of these, the cheapest and least harmful to tissue is 5 per cent trichloracetic acid, closely followed by 5 per cent formic acid. At least a hundred times the volume of the tissue is used. The solution is changed daily until decalcification is complete. The dilute acid solution should be prepared with distilled water.

End-point of decalcification

There are only two valid methods for determining whether decalcification of the tissue is complete:

1 X-raying of the tissue daily during decalcification.
2 Chemical testing of the used decalcifying solution at the time of daily changing.

Without doubt the X-ray method is better, but the apparatus is not always readily available; the majority must therefore use the chemical test. Bending the tissue, sticking needles into the tissue or cutting the tissue with a knife to check the absence of calcium are to be deplored.

Chemical test

Solutions required:

   (a) 0.88 s.g. ammonia
   (b) Saturated solution of ammonium oxalate

Technique

1 Take 5 ml of the decalcifying solution under test and place it in a clean test tube together with a piece of litmus paper.
2 Add ammonia drop by drop, agitating between each drop until reaction is alkaline. (If cloudiness or precipitate is present at this stage, decalcification is incomplete.)
3 Add 0.5 ml of saturated solution of ammonium oxalate and leave for 30 minutes. (If the solution remains clear, without precipitate or cloudiness, for this time, decalcification is complete.)

Note

Remember you are testing the solution and only indirectly the tissue. A faint precipitate or cloudiness indicates that the solution contains calcium but the tissue may be clear; so after putting the tissue in fresh solution repeat the test on this new solution in two hours.

DEHYDRATION

Dehydration: to remove water.

   If the selected tissue is to be processed to a non-water-miscible embedding media, then after fixation the tissue must be dehydrated.

The usual method is immersion of the tissue in ascending strengths of alcohol; for example, from 30 per cent rising in 10 per cent stages to 100 per cent, using the cheapest grade of alcohol compatible with the complete removal of water from the tissue.

The length of time the tissue spends in each strength of alcohol and the total time of dehydration depends on the tissue thickness, tissue density, and fixative used. If an alcoholic fixative has been used, then partial dehydration to a greater or lesser degree has already taken place and all that remains is to complete the process in up-graded alcohols.

On rare occasions, complete dehydration in absolute alcohol must be avoided; in this case arrest dehydration in 95 per cent alcohol and transfer tissue to aniline oil for 30 minutes, remove aniline oil with chloroform and 'clear' in the same reagent.

CLEARING

Clear: unclouded, transparent, not turbid.

Clearing may not be the most accurate term to use if the dictionary definition of 'clear' be applied; frequently the tissues do not become transparent, but translucent, depending on the type of tissue and reagent used.

The clearing reagent must usually be miscible with alcohol and wax, being the intermediary stage of processing where the clearing reagent replaces the alcohol and in turn is replaced by the paraffin wax.

Some reagents that could be used for clearing may be carcinogenic or toxic. So far as is possible the use of these should be avoided.

Xylene is the commonest, being quick, cheap and efficient.

Paraffin oil can be used.

Chloroform, although frequently used, is toxic, expensive and difficult to remove from the tissue. If removal is incomplete, the block will crumble.

Cedarwood oil is claimed to be the best but is little used, being expensive and messy. Except in specialised instances its handling and cost outweigh its advantages.

As with dehydration, the thickness of the tissue influences the length of time required to clear the tissue.

IMPREGNATION

Impregnation: to fill or saturate.

Impregnation of the tissue with paraffin wax, gelatine, celloidin or plastics is to replace the previous processing solution with a tissue-supporting medium. The length of time this processing stage takes depends on the viscosity of the impregnating medium being used and the density of the tissue.

For paraffin wax processing the following considerations influence the time factor, the clearing agent used and the use or non-use of a vacuum embedding bath.

EMBEDDING

Embedding: fix firmly in surrounding mass.

Embedding of the tissue consists of orientating the tissue in a molten impregnating medium that is then allowed to solidify.

METHOD FOR EMBEDDING TISSUE
IN PARAFFIN WAX

1 Fix tissue.
2 Select piece(s) of tissue for sectioning.
3 Decalcify tissue if required.
4 Dehydrate tissue up to 5 mm thick (thicker pieces will take longer).

    (a)   50 per cent alcohol   2 hours

    (b)   70 per cent alcohol   2 hours

    (c)   90 per cent alcohol   2 hours   See note 1

    (d) 100 per cent alcohol   2 hours

    (e) 100 per cent alcohol   overnight

    (f)   Xylene $1\frac{1}{2}$ hours   See note 2

    (g) Xylene $1\frac{1}{2}$ hours

    (h) 56 °C molten paraffin wax $1\frac{1}{2}$ hours   See note 3

    (i)   56 °C molten paraffin wax $1\frac{1}{2}$ hours

    (j)   56 °C molten paraffin wax   1 hour

    (k) Embed in fresh filtered 56 °C melting-point molten paraffin wax

Notes

1 Experience will teach how the dehydration schedule can be varied to suit different tissue, with the proviso that the removal of water from the tissue must be complete.
2 The appearance of the tissue will show when clearing is complete; an area exhibiting a dense opacity indicating incomplete clearing and a milky opacity indicating inadequate dehydration.
3 It is assumed that the reader does not have access to a heated vacuum embedding bath, the use of which will alter the time required to impregnate the tissue, and be advantageous in the processing of some tissues, i.e. lung and brain.
4 The method given is only a guide, but tissue up to 5 mm thick will be adequately processed by it.
5 If an automatic tissue processor is available then the time schedule will be much less.
6 Slices of tissue up to 2 mm thick may be processed in a day.

METHOD FOR EMBEDDING TISSUE
IN GELATINE

Gelatine mixture

Gelatine (high quality)	25 gm
Phenol	1 gm
Distilled water	100 ml

Dissolve the phenol in the distilled water and heat to a temperature just sufficient to dissolve the gelatine. Filter through gauze at a maintained temperature of 37 C. The $12\frac{1}{2}$ per cent solution is made by diluting this mixture.

Technique
1 Wash the fixed tissue overnight in running water.
2 Impregnate the tissue in $12\frac{1}{2}$ per cent gelatine solution at 37 C for 24 hours.
3 Impregnate the tissue in 25 per cent gelatine solution at 37 C for 24 hours.
4 Embed in 25 per cent gelatine, allow to solidify completely.
5 Remove block from mould and trim embedding mass with a sharp knife. Place trimmed block in 10 per cent formalin for at least 24 hours. Store block in this solution until finished with.
6 Section by the freezing method.

Notes
1 The fixative must be removed from the tissue before commencing processing; coagulation will otherwise occur before impregnation is complete.
2 Gelatine to some extent inhibits freezing; the surrounding embedding mass should therefore be kept to a minimum.
3 It is usual to section tissue by the freezing method without embedding it, but embedding is necessary with friable material or fragments.

GENERAL REFERENCES

ANDERSON, J. *How to Stain the Nervous System*. Livingstone, UK.

ARMED FORCES INSTITUTE OF PATHOLOGY. *Manual of Histologic and Special Staining Technique*. USA.

BAKER, J. *Principles of Biological Micro-technique*. Methuen, UK.

BANCROFT, J.D. *Histochemical Technique*. Butterworth, UK.

BANCROFT, J.D., and STEVENS, A. *Histological Stains and their Diagnostic Uses*. Churchill/Livingstone, UK.

BLUM, F. *Zif. wiss. Mikr.* (1893), Bd **10** k 314.

CONN, H.J. *et al. Staining Procedures*. Williams and Wilkins, USA.

CONN, H.J. *Biological Stains*. Williams and Wilkins, USA.

COWDRY, E.V. *Laboratory Techniques*. Williams and Wilkins, USA.

CULLING, C.F.A. *Handbook of Histopathological Technique*. Butterworth, UK.

DISBREY, B.D., and RACK, J.H. *Histological Laboratory Methods*. Livingstone, UK.

DRURY, R.A.B., and WALLINGTON, E.A. *Carleton's Histological Technique*. Oxford University Press, UK.

EVERSON PEARSE, A.G. *Histochemistry*. Churchill, UK.

HAM, A.W., and LEESON, T.S. *Histology*. Lippincott, USA.

LILLIE, R.D. *Histopathologic Technique and Practical Histochemistry*. McGraw-Hill, USA.

MALLORY, F.B. *Pathological Technique*. Saunders, USA.

MCCLUNG JONES, R. *McClung's Handbook of Microscopical Technique*. Hafner Pubg. Co., USA.

SMITH, A. Tissue shrinkage caused by attachment of paraffin sections to slides, its effect on staining. *Stain Technology* (1962) **37**, 339–345. USA.

SMITH, A. *Biological dyes. The histologist and the manufacturer. Laboratory Equipment Digest* (1965), **3**. UK.

SMITH, A. The use of frozen sections in oral histology. *J. Med. Lab. Tech.* (1962), **19**, Nos. 1 and 2. UK.

SMITH, A. Some observations on Museum Technique *J. Med. Lab. Tech.* (1959), **16**, 2, 53.

SILVERTON, R.E., and ANDERSON, M.J. *Handbook of Medical Laboratory Formulae*. Butterworth, UK.

(Journal.) *Stain Technology*. Williams & Wilkins, USA.

STEEDMAN, H.F. *Section Cutting in Microscopy*. Blackwell, UK.

UNDERHILL, B.M.J. *J. Royal Microscop. Soc.* (1932), **52**, 113.

WALKER HALL, and HERXHEIMER, G. *Methods of Morbid Histology and Clinical Pathology*. Green and Sons, UK.

INDEX

Figures in light type refer to technique text page numbers.
Figures in **bold** type refer to illustrations caption numbers.

191